Aerospace Engineering: From the Ground Up

WORKBOOK

Benjamin J. Senson
James Madison Memorial High School
Madison, Wisconsin

DELMAR
CENGAGE Learning

Australia • Brazil • Japan • Korea • Mexico • Singapore • Spain • United Kingdom • United States

Aerospace Engineering: From the Ground Up Workbook
Benjamin J. Senson

Vice President, Editorial: Dave Garza

Director of Learning Solutions: Sandy Clark

Acquisitions Editor: Stacy Masucci

Managing Editor: Larry Main

Senior Product Manager: John Fisher

Vice President, Marketing: Jennifer Baker

Marketing Director: Deborah Yarnell

Senior Marketing Manager: Erin Brennan

Marketing Coordinator: Jillian Borden

Senior Production Director: Wendy Troeger

Production Manager: Mark Bernard

Content Project Manager: David Barnes

Senior Art Director: David Arsenault

Technology Project Manager: Joe Pliss

For product information and technology assistance, contact us at
Cengage Learning Customer & Sales Support, 1-800-354-9706
For permission to use material from this text or product,
submit all requests online at **www.cengage.com/permissions.**
Further permissions questions can be e-mailed to
permissionrequest@cengage.com

Library of Congress Control Number: 2011940505

ISBN-13: 978-1-4354-4754-7

ISBN-10: 1-4354-4754-9

Delmar
5 Maxwell Drive
Clifton Park, NY 12065-2919
USA

Cengage Learning is a leading provider of customized learning solutions with office locations around the globe, including Singapore, the United Kingdom, Australia, Mexico, Brazil, and Japan. Locate your local office at: **international.cengage.com/region**

Cengage Learning products are represented in Canada by Nelson Education, Ltd.

To learn more about Delmar, visit **www.cengage.com/delmar**

Purchase any of our products at your local college store or at our preferred online store **www.cengagebrain.com**

Notice to the Reader
Publisher does not warrant or guarantee any of the products described herein or perform any independent analysis in connection with any of the product information contained herein. Publisher does not assume, and expressly disclaims, any obligation to obtain and include information other than that provided to it by the manufacturer. The reader is expressly warned to consider and adopt all safety precautions that might be indicated by the activities described herein and to avoid all potential hazards. By following the instructions contained herein, the reader willingly assumes all risks in connection with such instructions. The publisher makes no representations or warranties of any kind, including but not limited to, the warranties of fitness for particular purpose or merchantability, nor are any such representations implied with respect to the material set forth herein, and the publisher takes no responsibility with respect to such material. The publisher shall not be liable for any special, consequential, or exemplary damages resulting, in whole or part, from the readers' use of, or reliance upon, this material.

Printed in the United States of America
3 4 5 6 7 8 9 25 24 23 22 21

Contents

CHAPTER 7 ASTRONAUTICS   121

CHAPTER 8 AEROSPACE PHYSIOLOGY   131

CHAPTER 9 MATERIAL SCIENCE   137

CHAPTER 10 REMOTE SYSTEM DESIGN   161

Preface

This workbook was developed to support *Aerospace Engineering: From the Ground Up* with real-world, hands-on activities that build basic skills for engineering design and provide opportunities to apply those skills in more challenging projects. The author has applied years of experience teaching Project Lead The Way's® Aerospace Engineering curriculum to produce a resource brimming with the following:

- Hands-on, directed design activities
- Math and physics support
- Sketching experience
- Open-ended design problems and projects to provide greater challenges

Ample sheets of blank Engineers Notebook paper, as well as orthographic and isometric grids are included for practice. As students complete the variety of activities in this workbook, they will be prompted to use these resources to develop their documentation skills.

Features of this Workbook

This text was developed to complement and support Project Lead The Way's Aerospace Engineering curriculum, and can be used to support any project-based course in aerospace engineering. The following features are built into each chapter to help students apply the design process to achieve productive results.

BACKGROUND

Background sections help students develop and review the knowledge they need to perform the activities that follow.

TIP SHEETS

Tip sheets alert students to common pitfalls and provide helpful hints and motivating anecdotes to smooth students' journey toward successful design solutions.

EXERCISE

At the core of this workbook are dozens of hands-on exercises that build essential skills in math, physics, sketching and drawing, brainstorming, and teamwork.

Problem Sets

Chapters include problem sets for additional practice at varying levels of difficulty.

CASE STUDIES

This workbook includes several case studies that show how engineering concepts are applied in the aerospace industry and can be used to complete the activities in this workbook.

Additional Support for Teachers

Answers and solutions to the exercises in this workbook are provided for instructors at www.cengagebrain.com. At the cengagebrain.com home page, please enter the *core text* ISBN for *Aerospace Engineering* in the search box at the top of the page. This will take you to the product page where this resource can be found.

Aerospace Engineering and Project Lead The Way, Inc.

This workbook is part of a series of learning solutions that resulted from a partnership forged between Delmar Cengage Learning and Project Lead The Way, Inc. in February 2006. As a nonprofit foundation that develops curriculum for engineering, Project Lead The Way, Inc. provides students with the rigorous, relevant, reality-based knowledge they need to pursue engineering or engineering technology programs in college.

The Project Lead The Way curriculum developers strive to make math and science relevant for students by building hands-on, real-world projects in each course. To support Project Lead The Way's curriculum goals and to support all teachers who want to develop project/problem-based programs in engineering and engineering technology, Delmar Cengage Learning is developing a complete series of texts to complement all of Project Lead The Way's nine courses:

- Gateway to Technology
- Introduction to Engineering Design
- Principles of Engineering
- Digital Electronics
- Aerospace Engineering
- Biotechnical Engineering
- Civil Engineering and Architecture
- Computer Integrated Manufacturing
- Engineering Design and Development

To learn more about Project Lead The Way's ongoing initiatives in middle school and high school, please visit www.pltw.org.

Acknowledgments

The authors and publisher wish to thank everyone who assisted in the development of the text, especially the reviewers and Master Teachers who provided valuable input. The author extends special thanks to Project Lead The Way Master Teacher Jasen Ritter for his reviews of the content, and to Marion Waldman of id8-TripleSSS Publishing Services for her help in developing the manuscript. And finally, the author thanks his family for their tireless support throughout this project.

CHAPTER 1
An Introduction to Aerospace Engineering Careers

Skills List

After completing the activities in this chapter, you should be able to:

- Identify aviation careers of personal interest

- Use Internet resources to gather career-specific information

- Conduct interviews to gather information from subject area experts

- Summarize findings from diverse sources in one coherent presentation

- Connect current school course work with future career goals

BACKGROUND

Exploring Aerospace Industry Careers

Thinking about a career in aviation commonly calls up the image of a pilot or flight attendant because these are the most common aviation-related careers we are exposed to when we access the world of aviation (Figure 1.1). However, the wide diversity of careers available in the aerospace setting provides opportunities for many individuals with varied interests to find rewarding careers in related careers. In this chapter of the workbook, you'll research and explore a number of career options to become familiar with the preparation required during your high school career to access aerospace-related career opportunities.

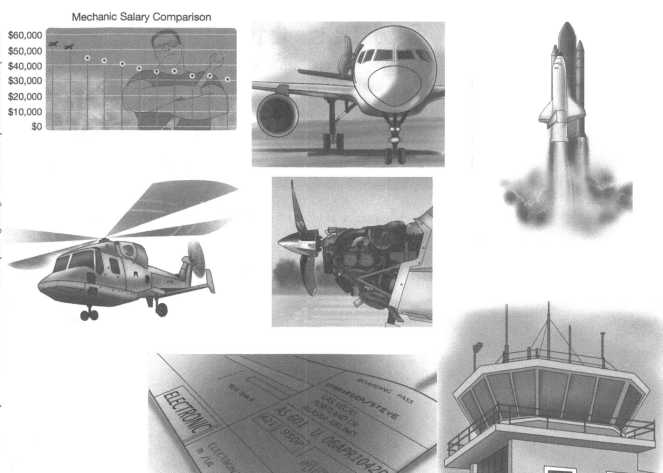

© Cengage Learning 2012

FIGURE 1.1 *The aerospace industry includes a wide range of occupations.*

Aviation careers . . . pilot in cockpit, mechanic on the engine, another repairs a wingtip, ground crew attaching tow tug, fuel truck nearby, tower controller in view, flight instructor teaching a class through window of building . . .

EXERCISE 1.1   AVIATION CAREER EXPLORATION

Objective

For this project, you'll work alone or in pairs to become the classroom expert on two aerospace-related careers. When you have completed this work, you'll be able to share your expertise with your classmates to produce an aerospace careers catalog that supports the career exploration efforts of other students in your school.

Materials

- ☐ Internet access
- ☐ Subject area experts
- ☐ Presentation and video editing software
- ☐ Poster board
- ☐ Engineer's notebook

Procedure

STEP 1 ▶ Selecting Careers to Explore

Your teacher will have a method for assigning you to specific careers, but for now look over the following list of aerospace-related jobs to identify a few that are of particular interest to you. Highlight or circle at least five of the listed occupations that are of interest to you based on their job title. Now use a different color of highlighter or pen to circle at least five of the listed occupations that you do not know what the work would look like based on the job title alone.

- Line Service Technician
- Electronics Technician
- Sheet Metal Mechanic
- Composites Technician
- Structural Mechanic
- A&P Mechanic
- Avionics Technician
- Aircraft Painter
- Interiors Mechanic
- Machinist
- Quality Control Inspector
- Munitions Technician
- Aeronautical Engineer
- Propulsion Engineer
- Mission Controller/Planner
- Flight Engineer
- Airport Manager
- Astronaut
- Flight Dispatcher
- Meteorologist
- Computer Engineer
- Computer Scientist
- UAV Pilot
- Test Pilot
- Charter Pilot
- Airline Pilot

- Cargo Pilot
- Med Flight Pilot
- Search and Rescue (Coast Guard) Pilot
- Military Aviator
- Load Master
- Flight Instructor
- FAA Inspector
- National Traffic Safety Board Investigator
- Airline Operations Supervisor
- Airport Manager
- Ticket Agent
- Baggage Handler
- Flight Attendant
- Transportation Safety Administration Agent
- Customs Agent
- Air Traffic Controller
- Aviation Safety Inspector
- Aircraft Cleaner

STEP 2 ▶ **What Do You Know?**

In small groups, carry out a brainstorming session to identify what you already think you know about the career choices that you have been assigned. Use a first word, last word approach to gather this information. In this method, one person is assigned to be the starter for the group. The starter simply states the title of one of their assigned occupations and then says nothing more. After the group briefly pauses to consider the job title, the person on the starter's right briefly shares his or her understanding of what the occupation involves and what a person in that career has to do to get a position in the occupation. All other group members remain quiet while they listen to the person speaking. After the first person finishes sharing, the right to speak passes to the next person in line until everyone in the group has shared their understanding of the career. This process does not involve questioning the statements of others; instead, everyone in the group simply shares information. The last word rests with the starter that first presented the career title. The starter's task is to summarize the statements of the whole group and then pass the starter role to the person to his or her right. The process is repeated until everyone has shared at least one of their assigned career choices.

STEP 3 ▶ **Exploring the Web**

Use Internet resources to access aviation career Web sites, job postings, postsecondary degree program descriptions, training guidelines, FAA regulations, industry growth predictions, and other relevant sites to gather information about your assigned careers. Use the space provided below to record your findings.

1. What are the job responsibilities and tasks associated with this career?

    _____

    _____

    _____

    _____

2. What are the working conditions for a person working in this career?

_____

_____

_____

3. Is postsecondary education or certification required to enter into this career? If so, what are the requirements of the training/educational programs?

_____

_____

_____

_____

4. What courses should you have taken by the time of high school graduation to be ready for admission into the training/certification programs required for this career?

_____

_____

_____

5. What is the current state of the career, and how is it expected to grow in the future?

_____

_____

_____

6. What is the starting and average pay rate for a person in this career area?

_____

_____

_____

STEP 4 **Accessing an Expert**

Although talking with one or two people will not give you an overview of an entire industry, it can provide meaningful insight into job satisfaction and the local wellness of an occupation. For this stage of your research, you should find a person actively involved in, or recently retired from, the occupation that you are researching. You might also consider contacting postsecondary training/certification organizations. Completing this phase of the project may involve setting up an interview in person or via email or phone, arranging for a job-shadowing day, or creating a survey to be completed by multiple people in the industry. Form your brainstorming group again to develop a series of 10 questions to ask during your access time with an expert. Record the 10 questions here:

1. _____

_____

2. _____

_____

3. _____

_____

4. _____
   _____

5. _____
   _____

6. _____
   _____

7. _____
   _____

8. _____
   _____

9. _____
   _____

10. _____
    _____

STEP 5 ▶ Reporting Your Findings

As a product of your research, you must produce one of the following resources and share it with the entire class. Support the work with a verbal introduction to the product, a summary of its contents and findings, and highlights of career characteristics that were surprising to discover.

Summary products:

- PowerPoint/keynote presentation
- Research poster session
- Promotional video
- Career pamphlet
- Other

STEP 6 ▶ Engagement, Motivation, and Inspiration . . . Bringing the Careers Home

Through the work that you completed during this assignment, you may have met a number of fascinating individuals involved with aerospace careers. As a class, have individuals nominate these career experts to be invited to speak to the whole class. If you are going to nominate someone, you should plan a brief 20-second speech that summarizes why you think this person will be of interest to most people in the class. After all the nominations are gathered, use whatever selection method you choose to narrow this list to two or three individuals, and invite them to speak to the class.

CHAPTER 2
History and Lighter-Than-Air Flight

Skills List

After completing the activities in this chapter, you should be able to:

- Calculate a net force as a product of buoyancy and weight

- Incorporate safety margins into product designs

- Know how altitude and temperature affect air density

- Understand the use of Commercial Off-the-Shelf (COTS) components in product development

- Quantify the variability in the quality of a product using standard deviation

- Apply standard volume calculations to simplify the design of complex vehicles

BACKGROUND

Understanding Lighter-Than-Air Vehicles

Since the time of Archimedes of Syracuse, we have understood that it is possible to manipulate and control the amount of buoyancy force that an object creates when it is immersed in a fluid. In the following activities and labs, you'll be asked to be the engineer, designing solutions using lighter-than-air methods to carry heavy loads up into the atmosphere (Figure 2-1).

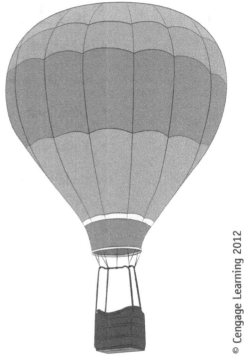

© Cengage Learning 2012

FIGURE 2-1 *A typical hot-air balloon is similar to the Cameron A-180.*

EXERCISE 2.1   IT'S A MATTER OF GAS

Objective

Understand how lift is created by lighter-than-air vehicles and the sequence of operations for launching a hot-air balloon.

CASE STUDY

A local hot-air balloon pilot has agreed to bring his balloon to your school site to offer students tethered ascents to a few 100 ft above the school's athletic field. The balloon is the "Buster One," which is a Cameron A-180 envelope with an Aristo-crat basket (Figure 2-2). This particular model has an envelope with a volume of 180,000 ft³ for the lifting gas.

On flight day, the balloon arrives packed in a trailer. You assist with the balloon setup procedure by helping to lay out large tarps on the ground on which you then spread out the balloon envelope. Next you haul a large basket, a burner unit, and a

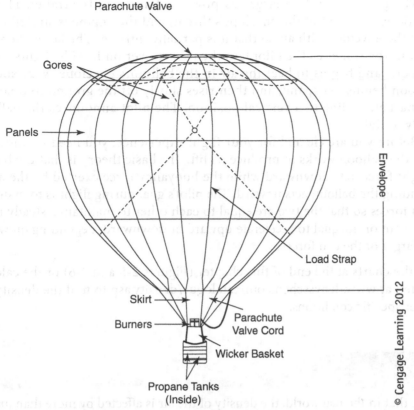

FIGURE 2-2 *Components of a typical hot-air balloon.*

propane gas tank from the trailer to near the envelope's open end. These are assembled by the pilot and then attached to the balloon's envelope by bundles of ropes that attach to load straps. The pilot explains that the load straps are sewn into the balloon and extend up to a small metal crown ring to which they are all attached at the top of the envelope. The pilot then attaches one end of a long stout rope to the balloon's basket and the other end to the trailer hitch of the truck that has been driven upwind of the launching site. This is the tether rope that limits the maximum altitude and range the balloon will reach during today's flights (Figure 2-3).

FIGURE 2-3 *Balloon tethered to a truck to restrict vertical flight.*

The pilot then rolls a large gas-powered fan out of the trailer. The fan is positioned in front of the envelope's throat, and the engine is started to cold-pack the envelope with air so that it is partially inflated. The fan is then stored back in the trailer as the pilot lays the basket over on its side, lights the twin burners, and begins to heat the air in the balloon's envelope. Very soon, the balloon begins to expand, and then rises slowly off the ground to float above the basket, pulling to a vertical position. The pilot announces the balloon is ready to fly!

Before you are cleared for your flight experience, you need to understand how the balloon works to produce its lift. The basic theory is that the balloon's weight force acts downward while the buoyancy force exerted by the air surrounding the balloon acts upward. The pilot's goal during flight is to manipulate both forces so that the two are equal to each other to maintain a steady motion in the air, or unequal to accelerate upward or downward, depending on which is the larger of the two forces.

Use the charts at the end of this chapter (Figures 2-5 and 2-6) or the calculator located at www.denysschen.com/catalogue/density.asp to find the density of air under specific conditions.

TIP SHEET

Note that in the real world, the density of the air is affected by more than just temperature. To go above and beyond this exercise, you can explore how the moisture content of the air affects its density and therefore the lift that can be produced by a lighter-than-air craft.

Practice Problems

1. If the buoyancy force is simply the mass of the air displaced by the hot-air balloon's envelope, then it can be calculated by multiplying the density of the air surrounding the balloon by the volume of air displaced. What is the buoyancy force on the "Buster One" if the air temperature is 70°F?

2. The balloon as a system has a total weight that is equal to the weight of each component of the system, including the heated air that is inside the envelope (Figure 2-4). What is the total weight of the "Buster One" given the following conditions?

Cameron A-180 envelope	=	368 lbs
Aristocrat basket (48" × 72")	=	210 lbs
Sirocco burner unit	=	54 lbs
3 propane tanks with full fuel (20 gal)	=	135 lbs each
Passengers (6 students)	=	145 lbs each
Pilot	=	185 lbs
Heated air (140°F average) inside the envelope	=	_____?

FIGURE 2-4 *A thermograph of a typical hot-air balloon shows the range in temperature inside the envelope.*

© Cengage Learning 2012

3. What is the net force on the fully loaded balloon under these conditions?

4. The pilot opens the blast valve on the burner to warm the air in the envelope just enough so that the net force on the balloon is zero, and it is neutrally buoyant. Using a wireless temperature sensor, the pilot can check how warm the air is in the envelope. What is the air temperature in the envelope under these conditions?

5. To ascend to the maximum altitude possible, the pilot holds the blast valve open until the average temperature in the envelope increases to 240°F. How strong must the tether rope tension be to restrain the balloon? This is a pretty important piece of safety equipment so incorporate a safety margin of 5:1.

6. To descend, the balloon needs to be cooled either by venting some of the hot air or by waiting for the air in the envelope to cool by conduction and radiation. How cold must the air in the envelope become to start the descent?

7. If the maximum average temperature that can be used in the envelope is 250°F, what is the maximum altitude to which the balloon can ascend if loaded like the "Buster One"?

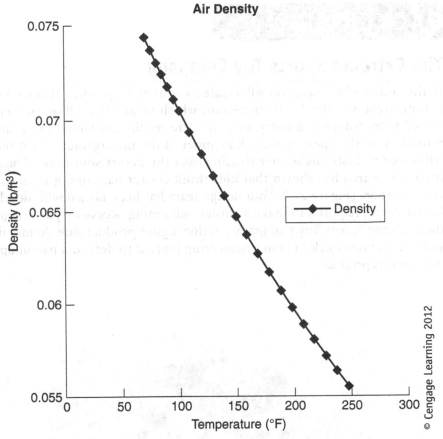

FIGURE 2-5 *Variation in air density as a function of temperature.*

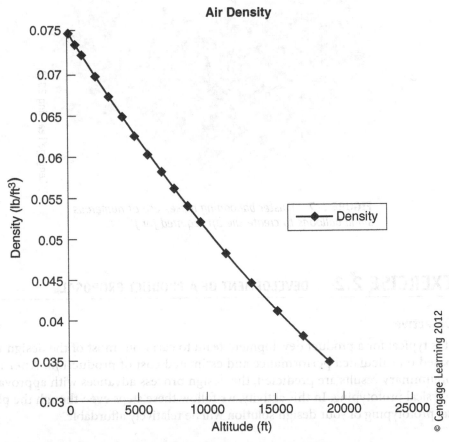

FIGURE 2-6 *Variation in air density as a function of altitude.*

BACKGROUND

The Extreme Sports Toy Company

In this design challenge, you will create a commercial product that uses COTS (Commercial Off-the-Shelf) materials, which means that all of the items required to prototype and test your design are readily available from numerous providers on the open market. A member of the management team recently witnessed a cluster ballooner floating over the desert southwest. Follow-up market research has shown that kids think cluster ballooning is an exciting extreme sport (Figure 2-7). Your design team has been tasked with developing the hardware required to make a cluster ballooning accessory pack to go with the Extreme Sports Toy Company's action-figure product line. Remember to show all of your work in your engineering journal to defend a patent application for this product.

© Cengage Learning 2012

FIGURE 2-7 *Cluster ballooning makes use of numerous small balloons to create the lift required for flight.*

EXERCISE 2.2 DEVELOPMENT OF A PRODUCT PROPOSAL

Objective

It is typical for a product development team to carry out most of the design work based on calculated performance and estimated cost of production. After these preliminary results are predicted, the design process advances with approval for physical prototyping. In this activity, we follow these steps even though the physical prototyping of your design solution will be relatively affordable.

Materials

- ☐ Action figure
- ☐ Balloons
- ☐ Baking muffin cups
- ☐ Pennies
- ☐ Helium tank with filler valve
- ☐ Balloon ribbon
- ☐ Scotch® tape
- ☐ Electronic balance
- ☐ 5-gallon bucket
- ☐ Water
- ☐ 4-cup measuring cup

Procedure

STEP 1 Obtain an action figure from the company warehouse. Determine the weight of the action figure, and record this value on the Product Development Data Sheet.

STEP 2 Obtain standard latex helium balloons, ribbon, and a helium tank with a fill valve from a local supplier.

STEP 3 Determine both the average weight and the variation in the weight of a typical latex balloon by taking data on at least 10 balloons. Record the mean weight and the standard deviation in the weight on the data sheet.

Example 2.1   Standard Deviation

A standard deviation and mean value help to describe the amount of variation present in a measured quantity that has a normal bell curve distribution such as the random variation in the weight of a latex balloon. In most cases, we are content to find the mean value for a variable and then to use it in all future calculations. However, if we know that there is a large amount of variation in a value, then we can predict the likelihood of our design failing due to that variation. About 68% of all items fall within one standard deviation, 95% fall within two standard deviations, and 99.7% fall within three standard deviations of the mean value.

To find a standard deviation, follow these steps:

1. Collect data to create a population of numbers to analyze.
2. Find the mean (average) value of the group.
3. For each value, subtract the mean value from the data point value, and square the resulting number.
4. Sum all of these resulting squares.
5. Divide by the number of data points in the data set.
6. Take the square root of this value to find the population's standard deviation.

Example:

There are five balloons with weights of 8, 7, 6, 7, and 6 grams, respectively.

The average weight of these balloons is:

$$\frac{8 + 7 + 6 + 7 + 6}{5} = 6.8$$

Find the square of the difference for each data point:

$(8 - 6.8)^2 = 1.44$

$(7 - 6.8)^2 = 0.04$

$(6 - 6.8)^2 = 0.64$

$(7 - 6.8)^2 = 0.04$

$(6 - 6.8)^2 = 0.64$

Sum all of the squared differences:

$1.44 + 0.04 + 0.64 + 0.04 + 0.64 = 2.8$

Divide by the number of data points, and take a square root:

$$\sqrt{\frac{2.8}{5}} = 0.75$$

This is one standard deviation, so 68% of all balloons should weigh the mean ± the standard deviation or 6.8 g ± 0.75 g, and 95% should weigh the mean ± 2 standard deviation or 6.8 g ± 1.5 g. We would describe this data set as being highly variable. Are real balloons more reliable in weight than this?

STEP 4 Assuming a calibrated diameter of _____ inches for a fully inflated balloon, and a spherical shape, calculate the predicted lifting capacity of a single balloon. The volume of a sphere is $Volume = \frac{4}{3}\pi r^3$. Record your results on the data sheet.

STEP 5 Find the weight per unit length for the ribbon material that will be used for securing the balloons to the action figure. Record the result on the data sheet.

STEP 6 Based on the results in the previous steps, develop a product proposal that lists all of the materials required for a successful cluster ballooning action figure. Incorporate the need for the predicted lift to be 120% of the total weight to be lifted. Do not include the action figure itself.

STEP 7 Design a calibrated measuring device to ensure that your balloons are inflated to a uniform diameter of _____ inches. Use your calibration device to inflate one balloon to full size, and tie it off with a measured length of ribbon. On the free end of the ribbon, tie a paper baking cup to the ribbon, and place enough pennies into the baking cup to hold down the balloon.

STEP 8 Place the inflated balloon with its tie down weight onto a scale, release the balloon, and measure the resulting down force on the scale (Figure 2-8). Record the result on the data table.

STEP 9 Untie the balloon from the baking cup, and weigh the baking cup with its pennies.

FIGURE 2-8 *Balloon weighed down by pennies in a baking cup.*

Source: Courtesy of Ben Senson

STEP 10 ▶ Submerge the balloon in a 5-gallon bucket that contains just enough water so that a fully inflated balloon can just barely be completely submerged in the water. Note that the starting water level is marked on the bucket. Completely submerge your balloon under the water being careful to just barely hold the balloon beneath the surface without submerging parts of your hand (Figure 2-9). Use masking tape and a pen to mark the new higher level of the water. Remove the balloon, and add water a cup at a time until the water level rises to the new higher level. Record the amount of water displaced by the balloon in cups and in cubic feet on the data sheet.

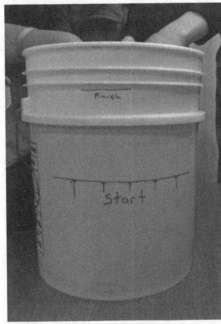

FIGURE 2-9 *Hold the balloon just below the surface of the water, and mark the volume change of the displaced water.*

Source: Courtesy of Ben Senson

STEP 11 Dry the balloon, and then release the helium (very small piece of Scotch tape and a pin works well), being careful to collect all of the pieces if it ruptures. Place the entire balloon and ribbon onto the scale, and record the resulting weight.

STEP 12 Based on your measurements, calculate the actual lifting force created by one fully inflated balloon. Record this value on the data sheet.

STEP 13 Summarize your research by using measured values to determine how many fully inflated balloons and what total length of ribbon is required to assemble a complete Action Figure Cluster Ballooning extreme sports accessory kit. When putting together your proposed solution to this design challenge, use the mean weight plus three standard deviations for the weight of a typical balloon. In addition, you will need to design your solution so that the lift produced is at least 120% of the actual weight to be lifted so the toy can rise into the air pulling a tether line with it. Double-check all your calculations prior to submission.

Product Development Data Sheet

Cluster Ballooning Development Project
Action Figure Accessory Kit - Design Analysis

Materials Characterization:
Action Figure:

Name of Figure Selected: _____

Weight of Action Figure: [ ]

Latex Balloons:

Individual Weights:

Mean Weight: [ ]

Square of Differences:
(Individual Weight – Mean Weight)²

Sum of Squared Differences: [ ]

Standard Deviation in Weight: [ ]

$$\sqrt{\frac{Sum\ of\ Squared\ Differences}{Number\ of\ Data\ Points}}$$

Weight Range:	1 standard deviation	[ ] +/–
	2 standard deviations	[ ] +/–
	3 standard deviations	[ ] +/–

Predicted Lift: Diameter = _____ [ ]

Lifting Capacity of Helium: 0.06295 lbs/ft³

Ribbon Material:

Total Weight: [ ]

Total Length: [ ]

Mass per Unit Length: [ ]

Cluster Ballooning Development Project
Action Figure Accessory Kit -
Parts Summary and Predicted Performance

Force Analysis:

Action Figure:

Weight of Action Figure:

☐

Latex Balloons:

Number Required:

☐

Total Weight:

☐ +/–

Total Predicted Lift:

☐

Ribbon Material:

Total Weight:

☐

Total Length Required:

☐

Total Weight of All Kit Components:

☐

Lift Required for Adequate Performance (F_{wt} x 120%)

☐

Sketch: Product and Packaging...

Cluster Ballooning Development Project
Action Figure Accessory Kit - Prototype Evaluation

Calibrated Balloon Analysis:
Latex Balloons:

Down Force Measured Inflated

Volume of Balloon (cups)

Volume of Balloon (ft³)

Weight of Baking Cup with Pennies

Wieght of Balloon with Ribbon

Actual Lifting Force per Balloon: (calculated)
Show All Work Here…

Product Parts Proposal:

Number of Balloons to be Used:

Total Length of Ribbon Required:

Total Weight of System with Action Figure:

Total Lift Produced:

Design Engineer's Signature: _____

Review Engineer's Signature: _____

BACKGROUND

Volume Matters!

The Monday staff meeting has quite a surprise for you and the rest of your designers. In the past, you have successfully designed lighter-than-air vehicles that are very traditional in design. Today, however, there is a new client who is interested in pushing the limits of design. Your design team has been tasked to think creatively and design a radically different lighter-than-air vehicle (Figure 2-10). To meet the client's specifications, the design must be capable of lifting a large party of at least 30 adults plus all of the typical balloon equipment. Assume a typical adult weighs 150 lbs and that the vehicle materials will weigh less than 1,000 lbs.

EXERCISE 2.3   PHASE ONE: DESIGNING FROM NORMS

Objective

Design a complex lighter-than-air vehicle shape.

Procedure

The marketing department believes that the design of custom vehicle shapes will be a very "hot" market for the foreseeable future. As such, they want your design team to build a basic library of shapes that can be scaled and integrated to create more complicated overall designs. This library should contain some basic geometric shapes (Figure 2-11). For each of the following design elements, calculate the total gas volume of the lifting envelope with the shape and dimensions indicated.

© Cengage Learning 2012

FIGURE 2-10 *Modern construction techniques allow balloons to have almost any imaginable shape.*

Sphere	Cylinder	Cone	Ellipsoid

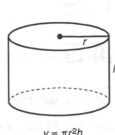

$v = \frac{4}{3}\pi r^3$

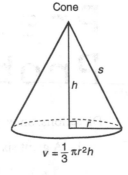

$v = \pi r^2 h$

$v = \frac{1}{3}\pi r^2 h$

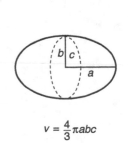

$v = \frac{4}{3}\pi abc$

Rectangular Prism	Isosceles Triangular Prism	Square-Based Pyramid

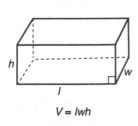

$V = lwh$

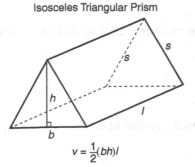

$v = \frac{1}{2}(bh)l$

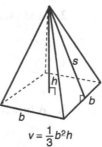

$v = \frac{1}{3}b^2 h$

FIGURE 2-11 *The volume of common geometric shapes.*

Example 2.2 Formulas for Volume (Simple Geometrical Shapes)

Volume of Sphere $= \left(\frac{4}{3}\right)\pi r^3$

Volume of Cylinder $= \pi r^2 h$

Volume of a Cone $= \left(\frac{1}{3}\right)\pi r^2 h$

Volume of Pyramid $= \left(\frac{1}{3}\right)b^2 h$

Volume of Cube $= s^3$

Volume of Rectangular Prism $= a\,b\,c$

Volume of Ellipsoid $= \left(\frac{4}{3}\right)\pi r_1 r_2 r_3$

Practice Problems

Directions: Show your work—formula, substitution, and solution!

1. A giant six-sided die measuring 40 ft on a side

2. A skyscraper that measures 20 ft by 30 ft at its base and is 100 ft tall

3. A bowling ball with a radius of 60 ft

4. A log that is 20 ft in radius on its end and 120 ft in length

5. A 1/10 scale model of the Great Pyramid of Giza measuring 76 ft on each side of its base and with a height of 48 ft

6. A giant vitamin E capsule that is 90 ft in its narrow dimension and 150 ft in length

7. The sugar cone from an ice cream cone (no scoop on top) with an opening at the top of 36 ft and a height of 96 ft

EXERCISE 2.3   (CONT.) PHASE TWO: A CUSTOM
                  MARKETING DESIGN

Procedure

Create a design for a lifting envelope that is a combination of at least three basic geometric shapes. Each shape can be used more than once, but at least three distinct geometric shapes must be used in the design. Make certain that the design is capable of providing the lift required to meet the specifications laid out by the marketing department. For this project, you will be using helium gas for the creation of lift rather than hot air. Helium has a lifting capacity of 0.06295 lbs/ft^3.

Use the design sketch form to show your dimensioned design. Include all values needed to predict the lifting capability of the balloon.

Use the mathematical analysis form to show all of the calculations required to predict the total lift capacity of your envelope design.

Design Title: _____ Designer's Name: _____ Date: _____

Design Sketch:

Designer Signature: _____ Date: _____ Witness Signature: _____ Date: _____

Design Title: _____ Designer's Name: _____ Date: _____

Mathematical Analysis:

Designer Signature: _____ Date: _____ Witness Signature: _____ Date: _____

BACKGROUND

The Great Balloon Hoax

On October 15, 2009, Richard and Mayumi Heene called 911 to report that an experimental balloon they were testing was accidentally launched with their almost 7-year-old son Falcon on board. The world watched for hours as the balloon ascended from the Fort Collins, CO backyard and drifted for almost 60 miles out into the desert before finally crashing to the ground with the boy nowhere to be found. Did he fall from the balloon? The world watched in nervous anticipation. Late in the afternoon, the boy was discovered hiding in the family garage and by the next day the entire episode had been revealed as a hoax. The question is, could we have known that the boy was not on the balloon? Using the following information, make a case for or against believing that the boy was onboard the balloon.

EXERCISE 2.4   CALCULATING THE LIFTING CAPACITY OF THE HEENE BALLOON

Objective

Use science and engineering principles to analyze a real-world case.

Materials

☐ Internet access

☐ Calculator

Procedure

> STEP 1   View the video of the accidental launch at www.youtube.com/ watch?v=kP_Cjx2sFzM. You may also Google other versions of the story to see additional images and coverage of the story. Based on the dimensions provided by Richard in his 911 call, the balloon measures 20 ft in diameter and 7 ft in height, including the gondola. Do these dimensions seem reasonable based on objects you can see in the field of view? What are the dimensions of the gas envelope without the gondola?

> STEP 2   Use the dimensions and the volume formulas to calculate the maximum lifting force that this balloon was capable of creating (Figure 2-12). What formula would give the best estimate given the shape of the balloon's gas envelope? Use the lifting force of 0.06295 lbs/ft³ for helium assuming that it was pure welding grade helium (much of the helium used to fill balloons is not this pure).

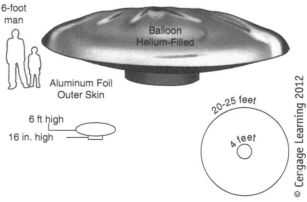

FIGURE 2-12 *Dimensions for the Heene family balloon.*

STEP 3 Find the total weight of the vehicle by adding the weight of Falcon and the vehicle together. For Falcon's weight, you would have had to consider how much a typical nearly 7-year-old child weighs by consulting the standard growth chart. One version is available through the Center for Disease Control at www.cdc.gov/growthcharts/data/set1clinical/cj41l021.pdf. Richard had stated that the vehicle only weighed 10 lb. What is the total weight to be lifted?

STEP 4 Fort Collins is located almost 5,000 ft above sea level, and the balloon ascended to nearly 8,000 ft above sea level during its flight. The density of the atmosphere rapidly decreases with altitude, which reduces the buoyancy force of a lighter-than-air vehicle. Use the following table to find the amount of lift produced by this balloon on the ground in Fort Collins and the maximum height to which it could have ascended (where lift equals total weight).

Altitude (Feet)	Pressure (P.S.I.)	Percentage of Pressure at Sea Level (%)
0, Sea Level	14.70	100
1000	14.43	98.2
2000	13.66	92.9
3000	13.17	89.6
4000	12.69	86.3
5000	12.23	83.2
6000	11.78	80.1
7000	11.34	77.1
8000	10.91	74.2
9000	10.50	71.4
10000	10.10	68.7
11000	9.8	67
12000	9.6	65
13000	9.3	63
14000	9.0	61
15000	8.3	58

STEP 5 Based on your calculations, was it predictable that Falcon was not onboard the balloon? Prepare a short presentation to summarize your evidence. Draw a conclusion and make the recommendation to approve or deny a request to begin a search-and-rescue operation.

CHAPTER 3
Basic Aerodynamics

Skills List

After completing the activities in this chapter, you should be able to:

- Calculate the lift and drag produced by a standard airfoil shape

- Use spreadsheet software to complete basic mathematical calculations

- Understand the changing nature of lift and drag during a typical flight profile

- Incorporate 3D modeling software such as Inventor to prepare for either 3D printing or creation of cutting templates

- Experience the use of hotwire manufacturing to create an airfoil shape

- Use a wind tunnel to gather flight-performance data

- Analyze the suitability of an airfoil profile to meet the constraints of a real-world application

Predicting and Measuring Flight Performance

Engineers and scientists frequently use basic mathematics to predict the performance of a design, scale an object to enlarge or reduce its dimensions, visualize patterns in data collected, analyze the amount of error present in a prediction or measurement, and consider the suitability of a solution for a particular set of design constraints.

In this unit, you will gain direct experience with the aeronautical engineering tasks of designing, predicting the performance of, building, testing, and then analyzing the performance of an airfoil for use in the creation of an actual aircraft. You will use industry-standard software tools such as spreadsheets to carry out mathematical calculations and 3D modeling software to transform your design geometry into an actual shape. You will also gain experience with one manufacturing technique for producing wing panels. Finally, we will use a wind tunnel to gather data on the lift and drag characteristics of your airfoil so that our final decision to approve or reject it for use in the final product is based on solid engineering research and data rather than opinion and conjecture.

EXERCISE 3.1   ANALYZING LIFT AND DRAG

Objective

Complete the lift and drag equations.

The lift and drag equations are very similar to each other and differ only in switching out the coefficient of lift with the coefficient of drag for a given flight condition. The lift equation is as follows:

$$Lift = C_L \times \rho \times A \times \frac{V^2}{2}$$

Complete this table:

Symbol	Variable	Unit of Measure (English System)	Unit of Measure (Metric System)
Lift			
C_L			
ρ			
A			
V			

Common equivalences:

$$1 \text{ N} = 1 \text{ kg m/s}^2$$

$$1 \text{ m} = 3.28 \text{ ft}$$

$$1 \text{ mph} = 1.4667 \text{ ft/sec} = 0.447 \text{ m/s}$$

$$1 \text{ slug/ft}^3 = 515 \text{ kg/m}^3 = 32.17 \text{ lb/ft}^3$$

BACKGROUND

The Aerial Medicine Show

In many locations, it is very challenging to provide medical care to individuals due to the scarcity of medical personnel and supplies, the remote location of the patient, or the need for immediate transportation to another location. In this exercise, you will consider the changes in lift and drag that occur during a fairly typical flight profile for a medical service provider in a remote part of the world (Figure 3-1).

© Cengage Learning 2012

FIGURE 3-1   *Aerial medical service flights over unpopulated areas.*

Today you are the mission coordinator for medical flights that will serve a number of patients in the area. Your company recently modified a 1966 Cessna 182 to serve as your air ambulance.

EXERCISE 3.2   A TYPICAL DAY OF RURAL MEDICINE FLIGHTS

Objective

Use the lift and drag equations to analyze an aircraft's flight performance during a typical flight.

TIP SHEET

The Cessna 182 has a planform wing area of 174 ft^2 (16.2 m^2), and the medical services airstrip is located at an altitude of 2000 ft MSL with a local current air density of 0.072 lb/ft^3 or 0.00223 slugs/ft^3 or 1.154 kg/m^3.

Practice Problems

The aircraft has a factory rated maximum weight of 2,800 lbs. After removing the rear seat to make room for a stretcher, the aircraft has an empty plus full oil weight of 1,622 lbs. Today's mission profile includes departure with a pilot/EMT (145 lbs), a doctor (110 lbs), and a physician's assistant (112 lbs), plus medical supplies (380 lbs) and a portable X-ray machine (75 lbs) on board. The aircraft currently has a full fuel load of 84 gallons (6 lbs/gallon).

1. First, find the total current weight of the aircraft. Second, find how many pounds of medical supplies or how many gallons of fuel need to be removed to comply with the factory takeoff weight limit of 2,800 lbs? (Round the gallons of fuel up to the next nearest gallon.)

2. For today's mission, the total flight distances are not anywhere close to the maximum range of the aircraft, but the medical supplies are essential for patients. Accordingly, you order fuel removed to meet the takeoff weight limit of the Cessna 182. After the fuel is pumped out and the supplies are fully loaded, the aircraft taxies to the runway. After a takeoff roll of 420 ft, the aircraft begins a steady climb at 88 mph indicated airspeed.

   During the climb, what is the C_L of the Cessna 182's NACA 2412 airfoil?

   Show all your work by writing the formula, rewriting it with the correct numbers and units substituted, and then providing a final solution.

TIP SHEET

Calculators are all designed to help you accurately solve mathematical problems; however, they all have unique characteristics that you need to get used to using. Common functions that you will need to solve these problems include finding the key or method for finding the square of a value, inverting a result, entering a number in scientific notation, or interpreting a solution that is already in scientific notation. Work with your teacher and other students to find out how to use your calculator, and then bring it with you on a daily basis.

3. The C_D is a sum of two major components of drag. The first is the parasitic drag caused simply by the net movement of the aircraft through the air which is described as C_{D0}. The second is the induced drag caused by the production of lift, which is described as C_{Di}. The induced drag is related to the coefficient of lift C_L by the following equation:

$$C_{Di} = \frac{C_L^2}{\pi \cdot AR \cdot e}$$

For the Cessna 182, the aspect ratio (AR) is equal to 7.366, planform efficiency factor (*e*) is 0.8, and the C_{D0} is equal to 0.025.

Find the amount of thrust required to maintain 88 mph during the climb. Use the density of the air at 2000 ft of altitude for this calculation.

4. After a while, the pilot transitions to cruising flight by reducing the pitch attitude of the aircraft and reducing throttle so that the aircraft travels at 129 mph at an altitude of 10,000 ft MSL where the air density is 0.00175 slugs/ft^3.

Find the new C_L for these flight conditions.

5. With the power adjusted to its 75% cruise setting, the aircraft maintains its speed. In this configuration, the propeller produces 290 lbs of thrust. What is the C_D during this cruising flight?

6. Find the coefficient of induced drag (C_{Di}) while in cruising flight.

7. The Cessna 182 burns approximately 9 gallons of fuel per hour of cruising flight. Four hours into the flight, you need to climb to 18,000 MSL to clear a mountain range. Once established at altitude, you level off in your original cruising attitude and configuration.

   What is the new velocity required to maintain altitude assuming air density is 0.00135 slugs?

8. Immediately after clearing the mountains, you land at a small city to drop off the doctor, supplies, and the X-ray machine. Before departure, you take a patient that weighs 68 lbs and 10 gallons of fuel on board. After a few hours delay waiting for some weather to clear, you take off and return to 18,000 MSL to head home. You find that you can hold your altitude with the same flight configuration and velocity as before.

   What is the new air density now that the weather system has passed?

9. Four hours later, the final stop for the mission drops the physician's assistant and the patient back in their original town. The recent rains in the area have made a mess of the grass strip runway. On takeoff, the aircraft's belly and landing gear is coated with 50 lbs of mud. At 10,000 MSL, you find that you can't reduce the throttle to hold altitude and maintain 129 mph. You know that full throttle thrust with this particular propeller is 350 lbs.

   What is the new C_D with the mud coating?

10. After safely returning from the mission, you realize that your future with the aeromedical services will require many landings on very small grass runways. To improve your safety, you want to calculate the actual stall speed that you need to maintain during approach to landing in order to land with minimal speed considering that the Cessna 182 has a maximum C_L of 1.4.

    What is the stall speed for an aircraft loaded only with a pilot and 10 gallons of fuel and one that is fully loaded to maximum gross weight?

BACKGROUND

Selecting an Airplane to Become a Radio-Controlled Model

As an engineer for Senson Tech Toy Company, you have been tasked with evaluating the flight performance of a radio-controlled (RC) scale-model aircraft. Your team will identify a real-world aircraft as an example (Figure 3-2).

Photo courtesy of Ben Senson

FIGURE 3-2 *The author poses with the Nextar RC aircraft.*

EXERCISE 3.3 DESCRIBING THE FULL SCALE AIRCRAFT

Objective

After you identify an aircraft, you will find its airfoil shape and obtain information on its predicted performance and the geometry of its 2D airfoil shape (Figure 3-3).

	A	B	C	D	E	F	G	H	I	J	K
1	NACA 2412 Airfoil as used on a Cessna 172 Skyhawk 1974 and Newer										
2											
3	α	[°]	0	5	10	15	20	25	30		
4	Cl	[-]	0.258	0.843	1.075	1.159	0.957	0.693	0.488		
5	Cd	[-]	0.01332	0.01641	0.0706	0.14005	0.25606	0.43742	0.69728		
6	Cm 0.25	[-]	-0.05	-0.056	-0.035	-0.027	-0.027	-0.028	-0.029		
7	Cp*	[-]	-0.574	-1.92	-5.864	-12.423	-21.334	-32.087	-44.355		
8	M cr.	[-]	0.695	0.494	0.316	0.226	0.174	0.143	0.122		
9											
10											
11	1	0									
12	0.999106	0.000205									
13	0.996177	0.000817									
14	0.991307	0.00183									
15	0.984515	0.003231									
16	0.975825	0.005004									
17	0.965269	0.007129									
18	0.952888	0.009581									
19	0.938727	0.012332									
20	0.922841	0.015354									
21	0.905287	0.018614									
22	0.886134	0.022078									
23	0.865454	0.025711									
24	0.843325	0.029476									
25	0.819834	0.033338									

Sheet1 | Raw Geometry and Performance | Sheet3

© Cengage Learning 2012

FIGURE 3-3 *Predicted lift and drag performance data and airfoil geometry information from JavaFoil.*

Screen capture by Ben Senson from Microsoft Excel

Airfoil test constraints:

- You will produce a wing test section with a 5 in. span, 3.5 in. chord, and the airfoil shape.

- Your wing section will be tested at angles of attack (AOA) of $-5°$, $0°$, $5°$, $10°$, $15°$, $20°$, $25°$, and $30°$, which should exceed the stall angle of any typical airfoil.

- Your wing section will be tested at speeds on a continuum but is evaluated at increments of 0, 10, 20, 30, 40, 50, and 60 mph.

- You have been employed to test whether or not a scale RC model of an aircraft (true scale requires the use of the original airfoil profile) with a 10.5 in. average chord can fly at your local fields.

- The full-size model will have a wingspan of 68 in. and will fly at 20 mph.

- The full-size model will weigh 7 lbs.

- Lift for the full-size model must exceed 200% of weight to be sufficient for beginner skill flight, 290% for intermediate flight, and 340% for advanced flight maneuvers.

- You will gather and summarize the evidence to support your recommendations to the management team of Senson Tech Toy Company regarding the feasibility of producing the proposed aircraft and the skill level recommendation in a technical report.

TIP SHEET

Reynolds Numbers and Flight Testing

A Reynolds number is a ratio between inertial forces and viscous forces in a fluid. For our purposes, the important characteristic of the Reynolds number is that it is proportional to both the velocity and the size of the object being tested.

$$RN = \rho VL$$

For the test of an object to be valid, it has to be tested at the same Reynolds number as that of the real-world situation. As such, if we test an object that is 10 times smaller than the real object, we need to test at a speed that is 10 times faster in order to have the same Reynolds number.

Procedure

STEP 1 ▸ Search for "The Incomplete Guide to Airfoil Usage" on the Web, and find a common or interesting aircraft that makes use of the same four- or five-digit NACA airfoil at both the root and tip of the wing.

STEP 2 ▸ Create a "NACA Decoder Key" for four- and five-digit NACA airfoils. Include labeled diagrams.

STEP 3 ▸ Write a design brief that summarizes the aircraft selected, the airfoil it makes use of, and how the aircraft is used in the real world. Include images or diagrams of the aircraft.

STEP 4 ▸ Investigate the airfoil's 2D geometry and predicted flight performance using the Web applet JavaFoil, and then transfer, process, and analyze this data in Excel.

STEP 5 ▸ Design, manufacture, and test a reduced-scale test model of the wing section.

TIP SHEET

Interpreting NACA Numbers

Four-digit series (e.g., NACA 2412):

- First digit is the camber of the airfoil as a percentage of the chord length. This airfoil has a 2% camber.

- Second digit is the location of the maximum camber point in tenths of the chord length as measured from the leading edge (LE) to trailing edge (TE). This airfoil's max camber is located at 4/10ths of the way from the LE to the TE of the airfoil shape.

- Third and fourth digits together represent the maximum thickness of the airfoil as a percentage of the chord length. This airfoil has a thickness of 12% of the chord length.

Five-digit series (e.g., NACA 23035):

- First digit times 0.15 gives the designed C_L or coefficient of lift. This airfoil has designed C_L of 0.3.

- Second and third digits together, divided by 2, indicate the distance from the LE to the maximum camber as a percentage of chord. This airfoils max camber is located 15% of the way from LE to TE.

- Fourth and fifth digits together represent the maximum thickness of the airfoil as a percentage of the chord length. This airfoil has a thickness of 35% of the chord length.

STEP 6 Start the JavaFoil applet, confirm you are on the Geometry tab, and choose the correct NACA series under Create an Airfoil.

STEP 7 Enter your NACA airfoil information, and adjust the Number of Points to equal 101.

STEP 8 Click on Create Airfoil.

STEP 9 Select the Velocity tab, and enter a range of AOAs from 0° to 30° in 5° steps. Scroll down and click on Analyze It!

STEP 10 Right-click in the data table, and select Copy.

STEP 11 Open Excel, and paste (Ctrl + V) into cell A10 of Sheet2. Leave Sheet1 blank.

STEP 12 Leave the information selected, copy it (Ctrl + C), select cell A1, and right-click to Paste Special, select Transpose, and click OK.

STEP 13 Delete the original pasted information.

STEP 14 Return to JavaFoil's Geometry tab, and right-click in the data window on the left side, click on Select All, right-click again, and click Copy.

STEP 15 In Excel, click on cell A10 of Sheet2, and paste the information in.

STEP 16 Click on the row 1 indicator tab, and insert two additional lines at the top of the spreadsheet.

STEP 17 Insert a title line in cell A1 such as "NACA #### as used for the ____," filling in the NACA number and aircraft name.

STEP 18 Right-click on the Sheet2 tab, and rename it "JavaFoil Geometry & Performance."

STEP 19 Save the Excel spreadsheet as "Yourname-NACA#."

BACKGROUND

Airfoil Scaling and Predicted Performance

An airfoil is a physical object that interacts with the flow of a fluid to produce lift forces. The actual shape of the airfoil and its alignment to the relative flow around it determines the strength of the lift and drag forces created.

The airfoil shape used for a particular aircraft design depends on the speed at which the aircraft will travel, the total force the wing needs to produce, and the necessity of being efficient.

To evaluate the flight performance characteristics of an airfoil, aerospace engineers commonly design and manufacture full-size or reduced-scale test objects that can be placed into a wind tunnel for directly measuring both the lift and drag forces produced by the shape over a broad range of speeds and AOAs.

EXERCISE 3.4 WING PROTOTYPING: SCALING FOR MANUFACTURE— TRANSFORMING DATA FROM JAVAFOIL WITH EXCEL

Objective

Scale your airfoil test object down to a size that can fit in the wind tunnel available for your use.

Procedure

STEP 1 Using the JavaFoil applet, you captured the geometry and predicted performance of a typical or common airfoil (see Exercise 3.3). Open this spreadsheet (Figure 3-4).

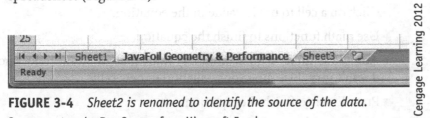

© Cengage Learning 2012

FIGURE 3-4 *Sheet2 is renamed to identify the source of the data.*

Screen capture by Ben Senson from Microsoft Excel

STEP 2 At the bottom of the screen, double-click or right-click the Sheet2 tab, and rename it "JavaFoil Geometry & Performance."

STEP 3 To scale the data in Excel, scroll down and note that the X coordinate data for the airfoil's surface starts with a value of 1 and then decreases to 0 before increasing back to 1. Note that the first and last set of X,Y data points are identical at 1,0. Highlight the last entry and right-click to Clear Contents to eliminate this duplication (Figure 3-5).

STEP 4 The JavaFoil geometry always gives a chord length of 1 from the LE to the TE. We need to scale the X and Y coordinates to fit the 3.5 in. chord length that we need for wind tunnel testing our airfoil.

STEP 5 To scale the coordinates, you can create two new columns labeled "Scaled X" and "Scaled Y" in your spreadsheet.

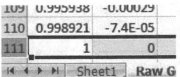

FIGURE 3-5 *Deletion of the final row of airfoil geometry data points to eliminate redundancy.*

Screen capture by Ben Senson from Microsoft Excel

© Cengage Learning 2012

STEP 6 In each column, you can enter a formula that multiplies all the X and Y coordinates by 3.5 (Figure 3-6).

	A	B	C	D	E	F	G	H	I	J
1	NACA 2412 Airfoil as used on a Cessna 172 Skyhawk 1974 and Newer									
2										
3	α	[°]	0	5	10	15	20	25	30	
4	Cl	[-]	0.258	0.843	1.075	1.159	0.957	0.693	0.488	
5	Cd	[-]	0.01332	0.01641	0.0706	0.14005	0.25606	0.43742	0.69728	
6	Cm 0.25	[-]	-0.05	-0.056	-0.035	-0.027	-0.027	-0.028	-0.029	
7	Cp*	[-]	-0.574	-1.92	-5.864	-12.423	-21.334	-32.087	-44.355	
8	M cr.	[-]	0.695	0.494	0.316	0.226	0.174	0.143	0.122	
9										
10	X	Y			Scaled X	Scaled Y		Scaling Multiplier		3
11	1	0			=A11*J10					
12	0.999106	0.000205								
13	0.996177	0.000817								

FIGURE 3-6 *Entering a formula into a cell starts with an equal sign. Note the use of $ symbols to lock the row and column.*

Screen capture by Ben Senson from Microsoft Excel

STEP 7 Label the original data columns as "X" and "Y," and then add labels to two new columns as "Scaled X" and "Scaled Y." Also create a label for a "Scaling Multiplier" cell.

STEP 8 The scaling multiplier will allow us to come back later and with a single change scale our airfoil up or down to match the constraints of a new wind tunnel or design project constraint.

STEP 9 Click on the cell where you want to enter the formula for the first "Scaled X" value.

STEP 10 Typing an equal sign (=) tells Excel that what follows is a formula.

STEP 11 Click on a cell to use its value in the equation.

STEP 12 Use math functions to finish the equation.

STEP 13 Add the symbol "$" in front of any value that you want to lock so it doesn't vary if we move our equation to new cells.

STEP 14 Press Enter to finish the equation.

STEP 15 Excel performs the calculation and shows the result in the cell.

STEP 16 Use Fill Down to transfer the formula into all of the remaining cells. To do this, select the cell with the formula, and hover the cursor over the black dot in the lower-right corner of the selection box. After it turns into a solid black cross, click and hold while pulling the box over the remaining cells. In this case, we want to pull it right one column under the Scaled Y label and downward until it does the math on all of our JavaFoil geometry data.

TIP SHEET

Using the fill tool normally causes the letter and number names of the cells used in an equation to change incrementally up one for every cell to the right (letter changes) and down (number changes). If you locked one or both of the cell name attributes, they will remain constant even as you "fill" the formula to new cells (Figure 3-7 and Figure 3-8).

	A	B	C	D	E	F	G	H	I	J
1	NACA 2412 Airfoil as used on a Cessna 172 Skyhawk 1974 and Newer									
2										
3	α	[°]	0	5	10	15	20	25	30	
4	Cl	[-]	0.258	0.843	1.075	1.159	0.957	0.693	0.488	
5	Cd	[-]	0.01332	0.01641	0.0706	0.14005	0.25606	0.43742	0.69728	
6	Cm 0.25	[-]	-0.05	-0.056	-0.035	-0.027	-0.027	-0.028	-0.029	
7	Cp*	[-]	-0.574	-1.92	-5.864	-12.423	-21.334	-32.087	-44.355	
8	M cr.	[-]	0.695	0.494	0.316	0.226	0.174	0.143	0.122	
9										
10	X	Y			Scaled X	Scaled Y		Scaling Multiplier		
11	1	0			3	0			3	
12	0.999106	0.000205			2.997319	0.000615				
13	0.996177	0.000817			2.988531	0.002452				
14	0.991307	0.00183			2.973922	0.005491				
15	0.984515	0.003231			2.953544	0.009694				
16	0.975825	0.005004			2.927474	0.015013				
17	0.965269	0.007129			2.895808	0.021386				
18	0.952888	0.009581			2.858664	0.028742				
19	0.938727	0.012332			2.816182	0.036997				
20	0.922841	0.015354			2.768522	0.046062				
21	0.905287	0.018614			2.715862	0.055842				
22	0.886134	0.022078			2.658402	0.066233				
23	0.865454	0.025711			2.596361	0.077132				
24	0.843325	0.029476			2.529976	0.088429				
25	0.819834	0.033338			2.459501	0.100015				
26	0.795069	0.03726			2.385207	0.111781				

⊮ ◀ ▶ ▶▮ JavaFoil Geometry & Performance ╱ Predicted Lift & Drag ╱ Actual Lift & Drag ╱ L vs D and ⁚

FIGURE 3-8 *Using a scaling multiplier in the formula allows you to change all of the calculated scaled values by changing one number.*

Screen capture by Ben Senson from Microsoft Excel

Upper Surface	
Scaled X	Scaled Y
-2.052	

FIGURE 3-7 *Results of an equation after entering the equation and pressing Enter.*

Screen capture by Ben Senson from Microsoft Excel

STEP 17 ▶ Print a copy of Sheet2, and add it to your engineering notebook or portfolio.

TIP SHEET

Note that if you use Print Preview in Excel, you can use Page Setup to tell it to Fit the printout to 1 page by 1 page or any set dimensions you want. You can also set Portrait versus landscape here. This can save a lot of paper, but just make sure it is still readable.

STEP 18 ▶ Switch to Sheet1 by clicking on its tab. In cell A1, enter the units of measure as "in" for inches and x, y, and z in cells A2, B2, and C2, respectively, as shown in Figure 3-9.

◢	A	B	C	D
1	in			
2	x	y	z	
3	3	0		
4	2.997319	0.000615		
5	2.988531	0.002452		
6	2.973922	0.005491		
7	2.953544	0.009694		
8	2.927474	0.015013		
9	2.895808	0.021386		

© Cengage Learning 2012

FIGURE 3-9 *Formatted for import into Inventor with units of measure, coordinates, and numerical data.*

Screen capture by Ben Senson from Microsoft Excel

STEP 19 In cell A3, you want to pull in the scaled geometry from the second sheet entitled "JavaFoil Geometry & Performance" but in a way that causes all of this information to automatically update if we later change the scaling factor.

STEP 20 Click on cell A3, and type in the "=" sign to start a formula.

STEP 21 Navigate to Sheet2, and then click on the cell of the first scaled X data value.

STEP 22 Press Enter.

STEP 23 Use the Fill feature to draw in all the data to Sheet1.

STEP 24 Rename Sheet1 to "Geometry for Inventor Import," and save the file again. Print out a copy for inclusion in your engineering notebook or portfolio (Figure 3-10).

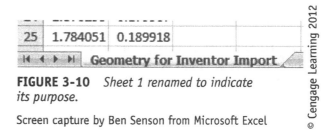

25	1.784051	0.189918	

ᴵ◀ ◀ ▶ ▶ᴵ **Geometry for Inventor Import**

© Cengage Learning 2012

FIGURE 3-10 *Sheet 1 renamed to indicate its purpose.*

Screen capture by Ben Senson from Microsoft Excel

STEP 25 Exit Excel.

BACKGROUND

Prototyping an Airfoil

In this project, you are introduced to one of two techniques for making physical models of an airfoil for testing purposes. The first procedure, using a hotwire cutter, will produce an airfoil out of a foam block. The second, optional method, using a 3D printer, will produce the same airfoil out of solid-surfaced (sparsely filled) plastic material.

EXERCISE 3.5   MODELING AND MANUFACTURING AN AIRFOIL

Objective

Manufacture the test object for wind tunnel evaluation by producing a reduced-scale wing section that makes use of your selected airfoil.

Materials

- Computer
- Access to JavaFoil applet
- Excel
- Word
- Inventor

If hotwire foam cutting airfoils:

- Scissors
- Spray adhesive
- Aluminum flashing material (0.015625 in. or thinner)
- Thumbtacks
- Hotwire foam cutter
- 6 × 5 × 2 in. thick foam (closed-cell construction grade, pink or blue)

If using a 3D printer:

- Printer
- Materials cartridges
- Support material cartridges

Procedure

Wing Prototyping: 3D Modeling

STEP 1   Using Autodesk Inventor software, you will create full-scale cutting templates for hotwire cutting a wing panel out of foam and/or a file to 3D print from the scaled airfoil data. Start by launching Inventor. Start a new project, rename it "Airfoil Design NACA####," and then click Finish (Figure 3-11).

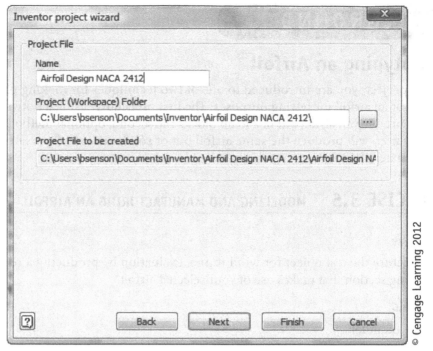

FIGURE 3-11   *Open and name a new project folder for your airfoil design work.*

Screen capture by Ben Senson from Inventor

Standard.ipt

FIGURE 3-12   *Icon for opening a new standard part.*

Screen capture by Ben Senson from Inventor

© Cengage Learning 2012

STEP 2 Start a new part file (Standard.ipt) (Figure 3-12).

STEP 3 Inventor should open in sketch mode; if it doesn't, start a new sketch.

STEP 4 Click on the Points tool to start the importing data points process. If you hover over any Inventor tool, a help dialog box opens and explains how the tool is used (Figure 3-13).

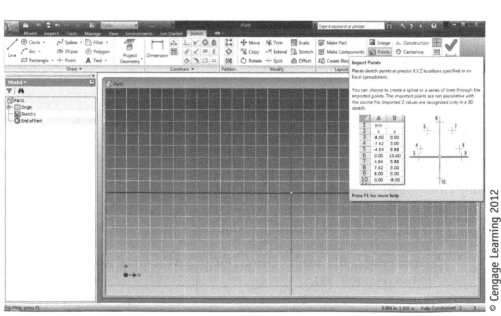

FIGURE 3-13   *From the Sketch menu, select the Points tool to import the Inventor data from Sheet1 of your spreadsheet.*

Screen capture by Ben Senson from Inventor

© Cengage Learning 2012

STEP 5 In the dialog box that opens, browse to find your Excel spreadsheet, and then select Create Points from the Options button at the bottom of the dialog box. Click OK. Note: Come back later and explore what happens if you use Create Spline instead ... don't worry, you can always Undo, so try both (Figure 3-14 and Figure 3-15).

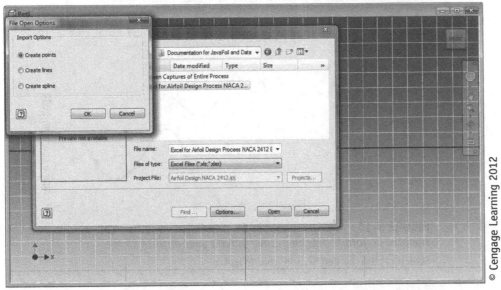

© Cengage Learning 2012

FIGURE 3-14 *In the Points Import dialog box, select Options, and use the Create Points option.*

Screen capture by Ben Senson from Inventor

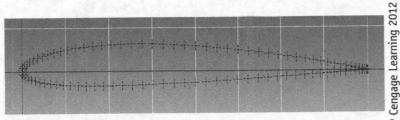

© Cengage Learning 2012

FIGURE 3-15 *Data points for the airfoils geometry after importing into Inventor.*

Screen capture by Ben Senson from Inventor

STEP 6 Select the Spline tool, and starting at the TE of the airfoil profile, click on every point on the upper profile, in order, stopping after clicking on the LE point. Shift-click exactly on a point if it is in error to remove it from the spline (Figure 3-16 and Figure 3-17).

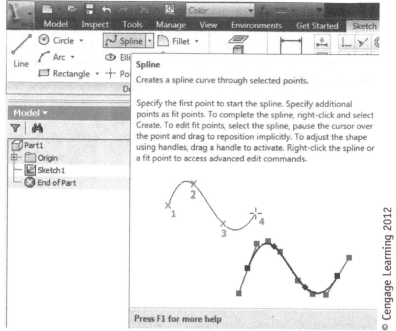

FIGURE 3-16 *The Spline tool in Inventor allows you to create a smooth, uniformly changing curve around your airfoil's shape.*

Screen capture by Ben Senson from Inventor

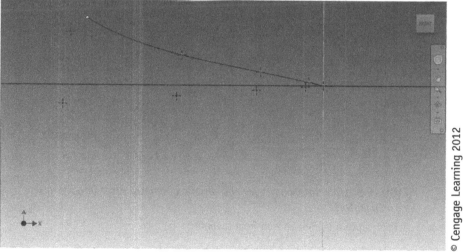

FIGURE 3-17 *Using the Spline tool, be certain to start at the airfoil's trailing edge and do not miss any data points.*

Screen capture by Ben Senson from Inventor

STEP 7 Right-click and select Create to finish the upper surface spline (Figure 3-18).

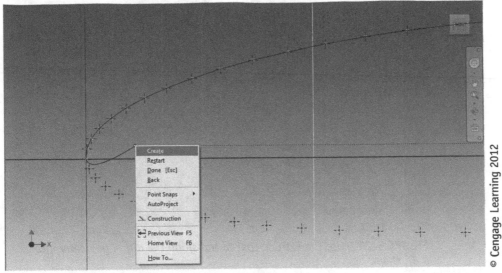

FIGURE 3-18 *At the airfoil's leading edge, right-click and choose Create to form the spline for the airfoil's upper surface.*

Screen capture by Ben Senson from Inventor

STEP 8 Repeat the process for the lower surface of the airfoil profile.

STEP 9 Use the Line tool to connect the LE data point to the TE data point on the airfoil profile (Figure 3-19 and 3-20).

Line

FIGURE 3-19 *Line tool selection icon.*

Screen capture by Ben Senson from Inventor

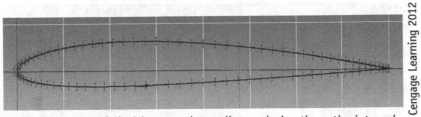

FIGURE 3-20 *Airfoil with a complete spline enclosing the entire internal structure of the airfoil.*

Screen capture by Ben Senson from Inventor

STEP 10 Finish the sketch.

STEP 11 Save the part with the name "NACA# Profile Only."

STEP 12 Save it again with the name "NACA# for 3D Printing."

STEP 13 Save it again with the name "NACA# for Hotwire Cutting All templates."

Preparation for 3D Printing

STEP 1 To add chord lines and embossed labels, open the part NACA# for 3D Printing.

STEP 2 Use the Extrude tool to transform the airfoil 2D profile into a 3D wing panel with a wingspan that matches your wind tunnel's capacity (5 in. for Jet stream 500). Click the profile of the airfoil, set to 5 in., and then click the OK button (Figure 3-21).

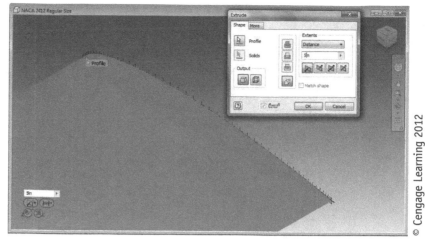

FIGURE 3-21 *Use the Extrude tool to transform the airfoil geometry into a 3D shape.*

Screen capture by Ben Senson from Microsoft Inventor

STEP 3 Apply a new sketch to the left end of the wing panel.

STEP 4 Use the Text tool to add "Yourname NACA#" to the end of the wing.

STEP 5 Finish the sketch, move the text to center it in the airfoil, and click to select it (Figure 3-22).

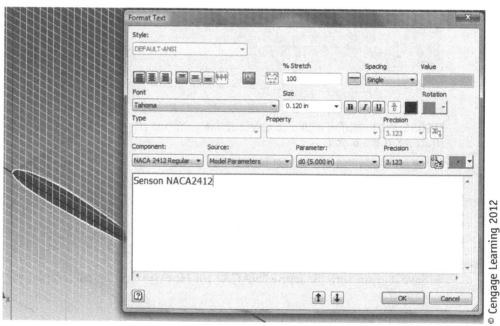

FIGURE 3-22 *Use the Text tool to place your name and airfoil number on a sketch on the wing panel's tip surface.*

Screen capture by Ben Senson from Inventor

STEP 6 Use the Emboss tool to inwardly emboss the text by 0.0625 in. (Figure 3-23).

FIGURE 3-23 *Emboss the text into the wing panel's flat end so it is a feature of the rapid prototype printed object.*

Screen capture by Ben Senson from Inventor

STEP 7 Rotate the wing panel, and add a new sketch to the right wing tip.

STEP 8 Use the Rectangle tool to draw a shape along the chord line (Figure 3-24).

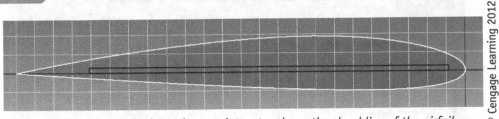

FIGURE 3-24 *Place a horizontal two-point rectangle on the chord line of the airfoil geometry on the wing panel tip without text.*

Screen capture by Ben Senson from Inventor

STEP 9 Finish the sketch, and extrude the rectangle into the panel by selecting the rectangular profile and a cut into the material of 0.125 in. Click OK (Figure 3-25).

FIGURE 3-25 *Extrude a cut with the rectangle's geometry 0.125 in. into the wing panel's tip so it prints as a feature of the printed object.*

Screen capture by Ben Senson from Inventor

STEP 10 Save the part.

STEP 11 Create a dimensioned drawing file with three views, and then print for your portfolio.

STEP 12 Render a view of the airfoil, and then print for your portfolio (Figure 3-26).

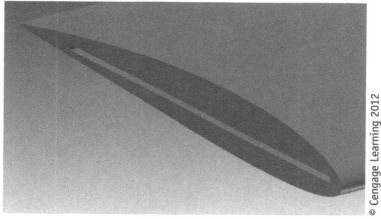

FIGURE 3-26 *Finished wing panel tip with chord line indicator feature.*

Screen capture by Ben Senson from Inventor

STEP 13 Save a copy as an STL file, and print this on the 3D printer (Figure 3-27 and Figure 3-28).

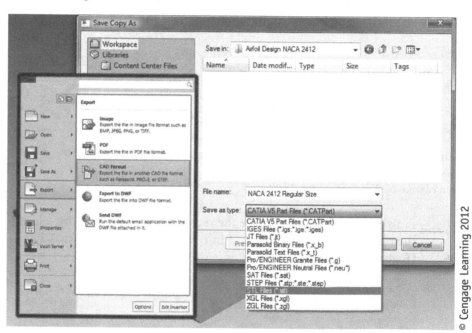

FIGURE 3-27 *Save your Inventor airfoil as an STL formatted file so it is ready for printing with a rapid prototype.*

Screen capture by Ben Senson from Inventor

FIGURE 3-28 *Completed airfoil shape printed as a 3D object.*

Rendered image by Ben Senson

Preparation for Hotwire Cutting

STEP 1 To create a cutting template, open the template named NACA# for Hotwire Cutting All Templates.

STEP 2 You need to create two cutting templates, one for each surface of the airfoil. To do this, you need to hook the LE and TE into place with a Fix constraint so they can't be moved. Select the Fix tool, and then click once on the LE and TE center points (Figure 3-29).

STEP 3 Use the Two Point Rectangle tool to draw a rectangle that represents the foam block around the airfoil shape. Click once up and to the left of the airfoil and then once down and to the right of the airfoil. Use the Dimension tool to make the rectangle 2 in. tall and 6 in. wide (Figure 3-30 and 3-31).

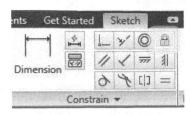

FIGURE 3-29 *Select the Fix constraint, and click on both the trailing and leading edge center points to lock their locations.*

Screen capture by Ben Senson from Inventor

© Cengage Learning 2012

⌐ Rectangle

FIGURE 3-30 *Two-Point Rectangle tool*

Screen capture by Ben Senson from Inventor

© Cengage Learning 2012

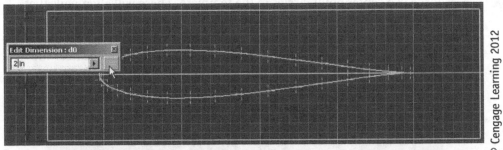

© Cengage Learning 2012

FIGURE 3-31 *Dimension the rectangle to represent the foam block out of which the airfoil will be cut.*

Screen capture by Ben Senson from Inventor

STEP 4 Finish dimensioning the foam block to place the TE 0.25 in. from the back and centered within the foam block. For very cambered airfoils, you may have to lock the airfoil off-center vertically to ensure that it fits easily within the block (Figure 3-32).

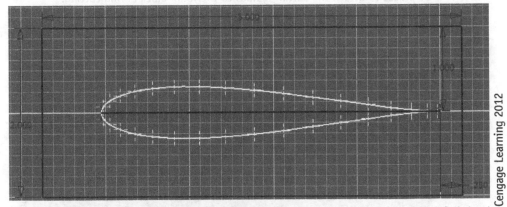

© Cengage Learning 2012

FIGURE 3-32 *Fully dimensioned "foam block" using dimensions to place the airfoil dead center in the block.*

Screen capture by Ben Senson from Inventor

STEP 5 Use the Line tool to create three lines that connect the TE to the center of the back of the foam block (wait for the green dots!), connect the LE to the upper-front corner, and connect the LE to the lower-front corner of the foam block (Figure 3-33).

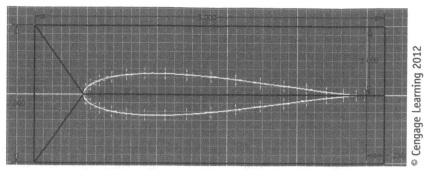

FIGURE 3-33 *Use the Line tool to place cutting guides from the leading edge to the corner of the foam block.*

Screen capture by Ben Senson from Inventor

STEP 6 ▶ Place two center point circles in the template. These will align nails that will be used to fasten the cutting template to the foam block. Place the two on the chord line for fairly symmetrical airfoils or along the mean camber curve for more cambered airfoils. One hole should be near the LE and one relatively far back in the airfoil. Dimension the circles to be 0.125 in. in diameter (Figure 3-34).

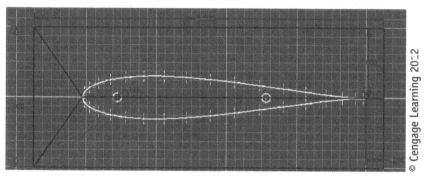

FIGURE 3-34 *Use center points and circles to indicate where pins can be placed to attach the cutting guides to the foam block during cutting.*

Screen capture by Ben Senson from Inventor

STEP 7 ▶ Finish the sketch by right-clicking away from objects in the sketch (Figure 3-35).

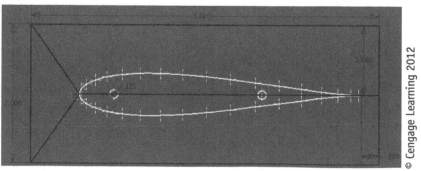

FIGURE 3-35 *Finished sketch for both the upper and lower cutting templates for hotwire cutting.*

Screen capture by Ben Senson from Inventor

STEP 8 ▶ Save a copy of the file twice—once as the "Upper Cutting Template" and once as the "Lower Cutting Template."

Extruding the Sketch to Create Cutting Templates

STEP 1 ▸ Close the file (don't save because you already have two copies), and then open the Upper Cutting Template file.

STEP 2 ▸ Use the Profile tool from the Extrude feature, and select the airfoil and bottom areas of the sketch for extrusion. Click on OK after they are selected (Figure 3-36 and Figure 3-37).

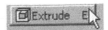

FIGURE 3-36 *Extrude tool icon.*

Screen capture by Ben Senson from Inventor

© Cengage Learning 2012

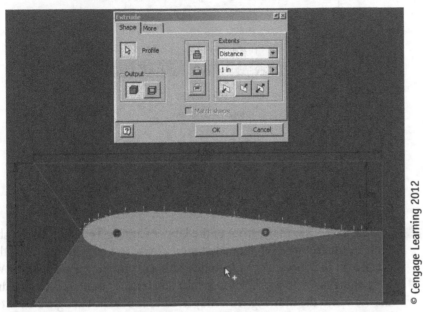

© Cengage Learning 2012

FIGURE 3-37 *Select the profile of the airfoil itself and one of the two cutting guide areas.*

Screen capture by Ben Senson from Inventor

STEP 3 ▸ Save the file (Figure 3-38).

© Cengage Learning 2012

FIGURE 3-38 *Final cutting template for guiding a hotwire over the upper surface of the airfoil when cutting foam block.*

Screen capture by Ben Senson from Inventor

Standard.idw

FIGURE 3-39 *Standard Drawing file icon.*

Screen capture by Ben Senson from Inventor

© Cengage Learning 2012

STEP 4 ▸ Open the Lower Cutting Template, extrude the airfoil and upper block, and then save the file. Close all open files.

Placing the Cutting Templates into a Drawing

STEP 1 ▸ Open a new Drawing file (Figure 3-39).

STEP 2 ▸ Right-click on Sheet1 in the lower-left browser window and choose Edit Sheet. Change the size of the sheet to an A size sheet, which is 8.5 × 11 in. paper (Figure 3-40).

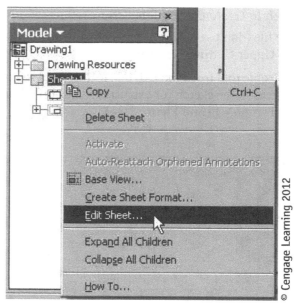

FIGURE 3-40 *The sheet size needs to be changed to match 8.5 × 11 in. standard letter size paper.*

Screen capture by Ben Senson from Inventor

STEP 3 ▸ Place a Base View of the front of the Upper Cutting Template on the sheet with a scale factor of 1:1. Note that you browse to find your file using the small icon that looks like a file folder with a magnifying glass over it. When everything is set, move the mouse into the drawing itself to click and drop the view into the drawing. You may need to move the dialog window to put the view where you want it. Press the Esc key on the keyboard after placing the base view, and then repeat to place a copy of the Lower Cutting Template into the drawing (Figures 3-41, 3-42, and 3-43).

FIGURE 3-41 *Base View selection tool.*

Screen capture by Ben Senson from Inventor

© Cengage Learning 2012

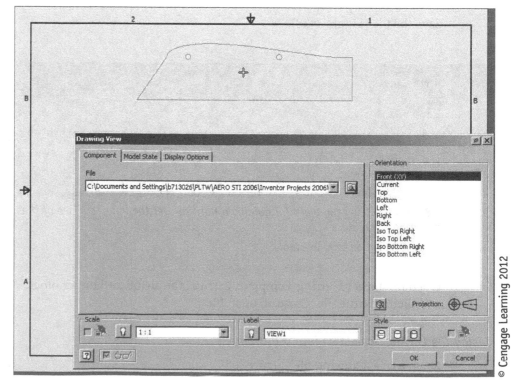

FIGURE 3-42 *Select and place both an upper and a lower template into the Drawing file at 1:1 scaling so they are full size.*

Screen capture by Ben Senson from Inventor

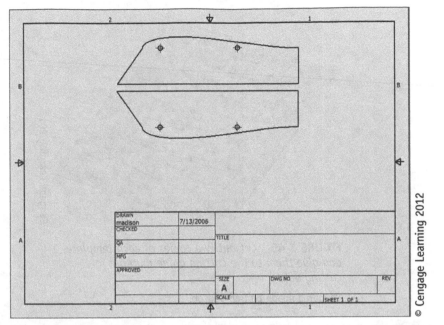

FIGURE 3-43 *Finished drawing file ready for printing.*

Screen capture by Ben Senson from Inventor

STEP 4 Click on the Annotate tab, select the Center Mark tool, and click on each of the circles visible in the drawing (Figure 3-44).

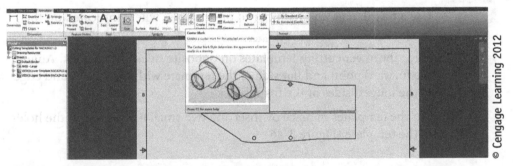

FIGURE 3-44 *Use the Annotation tool to place center points in each circle on the cutting templates.*

Screen capture by Ben Senson from Inventor

STEP 5 Save the file as "Cutting Templates."

STEP 6 Print three copies of the drawing file; two will be cut to make a template for *both ends* of each airfoil surface cut and the other needs to be placed in your engineering notebook.

Placing the Cutting Templates onto Sheet Aluminum

STEP 1 Trim the excess paper from two of the cutting template printouts.

STEP 2 Use spray adhesive to adhere the templates to a piece of sheet aluminum. Be very careful about sharp edges when working with sheet aluminum (Figure 3-45).

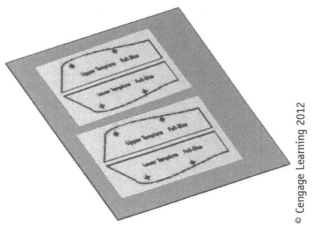

© Cengage Learning 2012

FIGURE 3-45 *Cut out two copies of each template, and glue them to the cutting guide material for cutting out.*

Rendered image by Ben Senson

STEP 3 Cut the templates from the aluminum sheet with scissors.

STEP 4 Using sandpaper or a fine file, smooth the edges of the newly cut templates. Rough edges snag on the hotwire causing safety hazards and rough surfaces on the airfoil. Also make sure that the similar templates are symmetrical.

Mounting the Cutting Templates on the Foam Block

STEP 1 Obtain a piece of 2 × 4 × 6 in. foam.

STEP 2 Use a board, nail, and hammer to punch holes where indicated on the templates.

STEP 3 Place the lower cutting templates on opposite sides of the block. Align the forward point and flat side of the template with the corner and edge of the longest sides of the foam block.

STEP 4 Pin the templates in place by inserting two small nails through the holes in the template (Figure 3-46).

© Cengage Learning 2012

FIGURE 3-46 *Foam block with a lower cutting template on both sides ready for hotwire cutting.*

Rendered image by Ben Senson

Hotwire Cutting

STEP 1 Turn on the hotwire power supply.

STEP 2 Perform a test cut on a piece of scrap foam. Adjust the wire temperature until the foam cuts at a steady rate and the wire's kerf is just wider than the wire diameter.

STEP 3 Begin your actual airfoil cut by guiding the wire along the straight section that leads into the TE of the air foil. In a fluid motion, continue guiding the foam over the wire until you have passed over the LE and out through the front of the block.

STEP 4 With the lower chord section cut, now change the templates. Remove the lower templates and replace them with the upper templates. When reinserting the pins, be sure to place them in the same holes used to fasten the lower templates. This will insure that the templates are aligned perfectly.

STEP 5 Complete the second cut to produce your finished airfoil.

STEP 6 Lightly sand the airfoil to smooth its surface (Figure 3-47 and Figure 3-48).

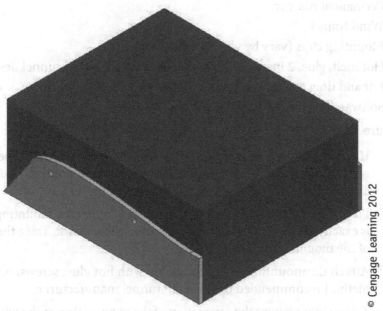

© Cengage Learning 2012

FIGURE 3-47 *Foam block with an upper cutting template on both sides ready for hotwire cutting.*

Rendered image by Ben Senson

© Cengage Learning 2012

FIGURE 3-48 *Final foam panel after both upper and lower cuts have been completed.*

Rendered image by Ben Senson

BACKGROUND

Wind Tunnel Testing

Wind tunnels have been used for over a century to gather flight performance data on new and existing designs. This data allows designers to have the proof of concept required to allow a new design to begin flight testing and can be invaluable for identifying failure modes of existing designs under controlled conditions.

EXERCISE 3.6   COLLECTING LIFT AND DRAG DATA

Objective

Gather lift and drag data points for wind speeds varying from 0 – 60 mph and AOAs varying from 0 – 30°.

Materials

- Airfoil prototyped wing panel
- Ruler
- Permanent marker
- Wind tunnel
- Mounting clips (vary by wind tunnel design)
- Hot melt, glue, 2 in. wide masking tape (varies by wind tunnel design)
- Lift and drag probes
- Software (control and data collection)

Procedure

STEP 1 Use the ruler to mark the bottom of the airfoil with a centerline that runs from LE to TE. On this line, mark the location of the "quarter chord" that is 25% of the way from the LE to the TE.

STEP 2 Place the mounting clip for the wind tunnel so that the mounting position is exactly above the centerline and quarter-chord mark. Trace the outline of the mounting clip onto the airfoil.

STEP 3 Attach the mounting clip to the airfoil with hot glue, screws, or other method recommended by the wind tunnel manufacturer.

STEP 4 Taking into account the orientation of the wing section in the wind tunnel, use the ruler and marker to draw in the chord line on the airfoil profile visible on the wing tip that will be toward the wind tunnel controller.

STEP 5 Mount the wing section in the wind tunnel at an AOA of 0°.

STEP 6 Collect lift and drag data for wind speeds ranging from 0 mph to 60 mph either continuously as wind speed increases or with stable wind conditions with steps of 5 mph.

TIP SHEET

Prepare to stop! As the drag on the wing section increases, it can start to vibrate or flutter on the mount. This can very rapidly lead to it breaking free from the mount, which can harm the wind tunnel. Learn how to stop the wind tunnel quickly, and be prepared to execute this rapid stop.

STEP 7 Adjust the AOA up to 30° repeating the experiment and collecting data at steps of 5°.

STEP 8 Save all data.

Flight Performance Analysis

Analyzing the data gathered from predictive models and from the wind tunnel testing of scale models often requires the use of mathematical tools such as spreadsheets. Spreadsheets speed up and automate many of the mathematical calculations required for a full analysis. Although setting up a spreadsheet takes forethought and attention to detail, the benefits are tremendous when constraints for a project are changed because with just a few adjustments, the entire analysis can be updated.

EXERCISE 3.7 EVALUATING THE MODEL AIRCRAFT DESIGN

Objective

Evaluate the ability to produce sufficient lift with minimal drag for a scale aircraft flown as an RC model.

Materials

- Data from Exercise 3.6
- Excel

Procedure

Calculating Predicted Lift and Drag

STEP 1 ▸ Open a new sheet in a spreadsheet such as Excel; rename it "Predicted Lift & Drag."

STEP 2 ▸ From Exercise 3.4, copy and paste the lift and drag coefficient information from cells A1 through 18 into the sheet.

STEP 3 ▸ Create additional columns to the right of the information copied for the remaining variables required to calculate the lift and drag produced by the airfoil. Note that you can get local air density from the national weather service and a local meteorologist. Otherwise, use an online tool such as "FoilSim" to calculate an approximate air density for your current altitude and temperature.

TIP SHEET

Aligning the Data!

Looking at the AOA data, we can see that it increases horizontally. To make use of Excel's ability to "fill" a table, we want to orient our velocity data so it increases downward.

Lift Equation: $L = C_L \cdot \rho \cdot A \cdot \frac{V^2}{2}$

Drag Equation: $D = C_D \cdot \rho \cdot A \cdot \frac{V^2}{2}$

STEP 4 Use Excel's text entering, shading, and border controls to format a data table for predicted lift as shown in Figure 3-49.

	A	B	C	D	E	F	G	H	I
1	NACA 2412 Airfoil as used on a Cessna 172 Skyhawk 1974 and Newer								
2									
3	α	[°]	0	5	10	15	20	25	30
4	Cl	[-]	0.258	0.843	1.075	1.159	0.957	0.693	0.488
5	Cd	[-]	0.01332	0.01641	0.0706	0.14005	0.25606	0.43742	0.69728
6	Cm 0.25	[-]	-0.05	-0.056	-0.035	-0.027	-0.027	-0.028	-0.029
7	Cp*	[-]	-0.574	-1.92	-5.864	-12.423	-21.334	-32.087	-44.355
8	M cr.	[-]	0.695	0.494	0.316	0.226	0.174	0.143	0.122
9									
10									
11	**Predicted Lift:**								
12									
13			AOA						
14			0	5	10	15	20	25	30
15	MPH	0							
16		10							
17		20							
18		30							
19		40							
20		50							
21		60							
22									

	J	K	L	M	N	O	P	Q	R	S	T
1		Additional Data for Calculating Lift and Drag:									
2											
3		Velocity (MPH):		**Velocity (ft/s):**			**Area (ft²):**			Wing:	
4		0		0.00				0.10417		Span	5
5		10		14.67						Chord	3
6		20		29.33			**Density (slugs):**			Area (in²	15
7		30		44.00				0.002			
8		40		58.67							
9		50		73.33							
10		60		88.00							

FIGURE 3-49 *Calculating predicted lift requires additional data such as velocities, density, and the wing panel's platform area.*

Screen capture by Ben Senson from Microsoft Excel

STEP 5 In the cell for AOA = 0 and mph = 0, enter the complete equation for calculating the lift produced by the airfoil under these conditions:

1. Click in the cell, and type an equal sign (=).
2. Click on the cell with the C_L for an AOA of 0°.
3. Type an "*" for multiplication.
4. Click on the cell with the air density in slugs.
5. Type an "*" for multiplication.
6. Click on the cell with the planform area of the wing section.
7. Type an "*" for multiplication.
8. Click on the cell with the velocity of 0 mph.
9. Type in an "*" for multiplication.
10. Type in "0.5," and press Enter.
11. Click in the cell with the equation entered into it, and highlight and modify the term in the equation for the velocity to format it to square the term. Note that to raise a term to a power, you press Shift + 6 for the ^ symbol. When you are done, it will look like this:(M4^2).
12. Press Enter. When you are done, the equation will look similar this: = C3*Q7*Q4*(M4^2)*0.5.

STEP 6 Click on the cell with the equation entered into it.

In the equation editor window, add a "$" symbol (without the quotes) before any variable in the equation that you don't want to change as you drag this equation horizontally (where the letters shift through the alphabet with each slide of one cell) and vertically (where the numbers increment by one for each cell shifted). For example, the cell Q7 contains the air density data. If this equation is pulled downward, it would increment to Q8 then Q9, and so on. However, if it is edited to be Q7 in the equation before the equation is filled, then it will remain locked on cell Q7. You can lock only the letter, only the number, or both through the use of the $ symbol.

STEP 7 Modify the equation to lock the appropriate variables so they can't change during a fill operation. The equation will look similar to this: $= C\$3*\$Q\$7*\$Q\$4*(\$M4^2)*0.5$.

STEP 8 Place the cursor over the lower-right corner of the equation box, and click and pull it vertically downward to fill the equation into all cells up to the maximum wind speed. Release the click.

STEP 9 With the entire column of 0° AOA cells still highlighted, click in the lower-right corner of the border, and hold to drag horizontally to the maximum AOA cell. Release the click.

STEP 10 The entire table should be filled with calculated lift performance data.

STEP 11 Repeat to create a similar Predicted Drag table below the lift table.

STEP 12 Highlight the entire Predicted Lift table, including the black box in the upper-left hand corner, but do not include the text labels outside of the table.

STEP 13 Use the Insert function to add a Chart, Surface, and 3D Surface.

STEP 14 Drag the chart to an appropriate location to the right of the table.

STEP 15 Repeat to create a surface chart of the Predicted Drag data.

Summarizing Actual Lift and Drag

STEP 1 Open a new sheet in the spreadsheet, and rename it "Actual Lift & Drag."

STEP 2 Create tables for Actual Lift and Actual Drag that are identical to the formatting used for Predicted Lift and Predicted Drag.

STEP 3 Insert your data from the wind tunnel experiments by importing all of the data into columns in the sheet and then copying and pasting the necessary data points into the table using those that most nearly coincide with the required wind speeds.

STEP 4 Insert a surface chart for both the Actual Lift and the Actual Drag tables.

Calculating Percent Error

STEP 1 Start a new sheet, and rename it "Percent Error."

STEP 2 Create new tables for Percent Error in Lift and Percent Error in Drag.

STEP 3 Use the previously learned techniques to create an equation in the cell for AOA and wind speed of 0 that can then be "filled" to the entire table.

TIP SHEET

Data from other sheets can be pulled into an equation simply by clicking on the tab for the sheet where the data can be found, clicking on the cell desired, and pressing the Enter key.

Percent Error:
$$Percent\ Error = \frac{|Estimated - Observed|}{Observed} \cdot 100\%$$

Evaluating Lift Produced

STEP 1 Open a new sheet, and rename it "Total Lift."

STEP 2 Create an equation that uses the actual lift produced by the wing section test object for flight conditions that most closely resemble typical cruising flight in both AOA and wind speed for the RC model aircraft scaled up to the full size of the model's wing planform size. Be certain to take Reynold's Number (RN) effects into account as you complete your analysis.

STEP 3 Consider whether this airfoil is capable of meeting the design requirements for the RC model selected in Exercise 3.3.

STEP 4 Write a technical report summarizing your research findings and concluding with recommendations for improvements to this design evaluation process as well as whether or not the aircraft you selected can successfully be manufactured as a true scale model at the size that was proposed.

TIP SHEET

Technical engineering reports typically contain the following sections:

- Abstract
- Table of Contents
- Introduction
- Background
- Materials
- Procedure
- Results
- Conclusions
- Appendix Materials

CHAPTER 4
Flight Dynamics

Skills List

After completing the activities in this chapter, you should be able to:

- Evaluate an aircraft's weight and balance in preparation for flight

- Incorporate safety margins into product designs

- Understand the use of mathematical, graphical, and table methods for solving a problem

BACKGROUND

Weight and Balance

All aircraft have a center of mass that represents the location at which we can act as if the entire vehicle exists. Aerodynamic and gravitational forces applied anywhere other than at the center of gravity tend to create rotational forces (moments) that cause the aircraft to rotate in space. Controlling the size and location of these forces is a critical part of maneuvering the aircraft in flight. Maintaining the balance of forces within limits is critical for ensuring that the vehicle remains controllable. As such, finding the weight and balance of the aircraft prior to the beginning of flight is a legal requirement for the pilot in command (Figure 4-1).

© Cengage Learning 2012

FIGURE 4-1 *Loading baggage into the cargo area of a small commuter aircraft.*

EXERCISE 4.1　CALCULATING MOMENTS

Objective

Become skillful at determining the weight and balance condition of an aircraft.

Example 4.1　Calculating Safe Load Weight

Determining whether an aircraft is safely loaded for flight can be accomplished by a number of different methods. The first of these is calculation by multiplying the weight of an object by the distance from its center of gravity to some datum line (arm). All of the major components for a typical aircraft have a known location as shown in Figure 4-2. The basic empty weight of the aircraft itself also has to be taken into account. This weight generally includes any unusable fuel as well as all permanent components of the aircraft.

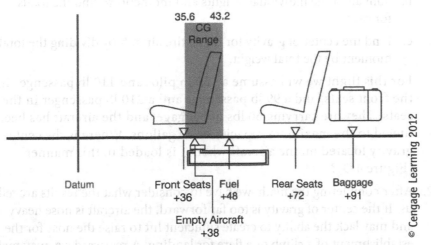

FIGURE 4-2 *Representative locations from the datum for the major items that have to be taken into account during a weight and balance preflight analysis for a small general aviation aircraft.*

Source: Based on *The Federal Aviation Handbook*, figure 2-15, www.faa.gov/ library/manuals/aircraft/media/FAA-H-8083-1A.pdf

Practice Problems

1. Use the following method to complete the chart provided in Figure 4-3.

   a. Solve for the moment of each basic component.

   b. Sum all of the individual weights and moments to find the totals for each.

   c. Find the center of gravity for the entire aircraft by dividing the total moment by the total weight.

   For this flight we will assume a 150 lb pilot and 110 lb passenger in the front seats, and a 99 lb passenger and a 210 lb passenger in the rear seats. They are carrying 60 lbs of baggage, and the aircraft has been fueled to the maximum capacity of 40 gallons. Where is the center of gravity located on the aircraft when it is loaded in this manner (Figure 4-3)?

2. After completing the math, we have to consider what the results are telling us. If the center of gravity is too far forward, the aircraft is nose heavy and may lack the ability to create sufficient lift to raise the nose for the establishment of a climb or a flare for landing. A rearward c.g. may make it impossible to prevent the nose from rising to prevent a stall. If this aircraft has a safe range for the center of gravity of 35.6 in. to 43.4 in., is the aircraft safe to fly?

3. If an airplane is overloaded, it can cause many problems. The aircraft will require an extended takeoff roll, climb more slowly, cruise more slowly, suffer from reduced range, be less maneuverable, require a faster landing

Weight and Balance Calculation

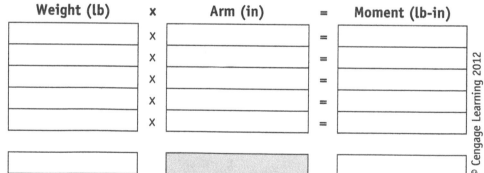

Item:	Weight (lb)	x	Arm (in)	=	Moment (lb-in)
Empty Weight of Aircraft		x		=	
Pilot and Front Seat Occupants		x		=	
Rear Seat Occupants		x		=	
Fuel (6 lbs/gallon)		x		=	
Baggage		x		=	
TOTALS					

FIGURE 4-3 *Typical table for table method of solving aircraft weight and balance problems.*

approach and longer landing roll, and will expose the aircraft structure to much higher loads. If the aircraft has a maximum allowable weight of 2,350 lbs, is the aircraft safe to fly?

4. You will find that the aircraft is suitable to fly based on total weight; however, the loads are distributed too far away from the datum point, which makes the aircraft tail heavy. Consider the following solutions in regards to whether or not they solve the problem of a rearward center of gravity. Which is the simplest and most useful solution? For each possible solution show all your work!

© Cengage Learning 2012

- **Method 1:** Leave behind the baggage.

- **Method 2:** Leave behind one passenger.

- **Method 3:** Have the fuel truck pump off 10 gallons of fuel.

- **Method 4:** Have the 110 lb and the 210 lb passengers switch seats.

Example 4.2   Using Tables to Look Up Moments for Components

A second method for analyzing the weight and balance condition of an aircraft is by using tables to look up the moments for the various components. An example is shown here for the fuel tanks. Note that the actual moments are 100 times larger than the number in the chart. This method is common for aircraft that have large moments (see Table 4-1).

Usable Fuel at Arm of 48"		
Gallons	Weight (lb)	Moment (lb-in/100)
5	30	14.4
10	60	28.8
15	90	43.2
20	120	57.6
25	150	72.0
30	180	86.4
35	210	100.8
40	240	115.2

TABLE 4-1   *Usable fuel at arm of 48".*

The chart shown in Figure 4-4 is typical for a large passenger aircraft. Notice that the chart builds in a safety factor by combining a given number of passengers sitting within a particular seating area.

PASSENGER LOADING TABLE

Number of Pass.	Weight lbs	Moment 1000
Forward Compartment Centroid — 582.0		
5	850	495
10	1,700	989
15	2,550	1,484
20	3,400	1,979
25	4,250	2,473
29	4,930	2,869
AFT Compartment Centroid — 1028.0		
10	1,700	1,748
20	3,400	3,495
30	5,100	5,243
40	6,800	6,990
50	8,500	8,738
60	10,200	10,486
70	11,900	12,233
80	13,600	13,980
90	15,300	15,728
100	17,000	17,476
110	18,700	19,223
120	20,400	20,971
133	22,610	23,243

CARGO LOADING TABLE

Weight lbs	Moment 1000 Forward Hold Arm 680.0	Aft Hold Arm 1166.0
6,000		6,966
5,000	3,400	5,830
4,000	2,720	4,664
3,000	2,040	3,498
2,000	1,360	2,332
1,000	680	1,166
900	612	1,049
800	544	933
700	476	816
600	408	700
500	340	583
400	272	466
300	204	350
200	136	233
100	68	117

FUEL LOADING TABLE

TANKS 1 & 3 (EACH)			TANKS 2 (3 CELL)					
Weight lbs	Arm	Moment 1000	Weight lbs	Arm	Moment 1000	Weight lbs	Arm	Moment 1000
8,500	992.1	8,433	8,500	917.5	7,799	22,500	914.5	20,576
9,000	993.0	8,937	9,000	917.2	8,255	23,000	914.5	21,034
9,500	993.9	9,442	9,500	917.0	8,711	23,500	914.4	21,488
10,000	994.7	9,947	10,000	916.8	9,168	24,000	914.3	21,943
10,500	995.4	10,451	10,500	916.6	9,624	24,500	914.3	22,400
11,000	996.1	10,957	11,000	916.5	10,082	25,000	914.2	22,855
11,500	996.8	11,463	11,500	916.3	10,537	25,500	914.2	23,312
12,000	997.5	11,970	12,000	916.1	10,993	26,000	914.1	23,767
FULL CAPACITY			**(See note at lower left)			26,500	914.1	24,244
						27,000	914.0	24,678
**Note: Computations for Tank 2 weights for 12,500 lbs to 18,000 lbs have been purposely omitted.			18,500	915.1	16,929	27,500	913.9	25,132
			19,000	915.0	17,385	28,000	913.9	25,589
			19,500	914.9	17,841	28,500	913.8	26,043
			20,000	914.9	18,298	29,000	913.7	26,497
			20,500	914.8	18,753	29,500	913.7	26,954
			21,000	914.7	19,209	30,000	913.6	27,408
			21,500	914.6	19,664			
			22,000	914.6	20,121	FULL CAPACITY		

FIGURE 4-4 *Loading table for a large commercial aircraft.*
Source: Based on *The Federal Aviation Handbook,* figure 7-2

Example 4.3 Using a Chart to Find Moments for Components

The third method for finding the moments caused by each component is to use a chart such as that shown here. Note that you enter the chart at the left side with the weight of the component, move horizontally until you reach the chart line for the component, and then move vertically down to read off the total moment for that component. Notice that this chart divides the total moment by 1000 (Figure 4-5).

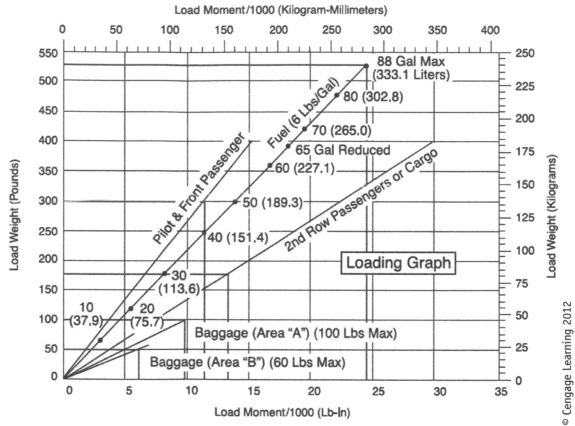

FIGURE 4-5 *Typical loading chart for a small four-seat general aviation aircraft.*

Source: *The Federal Aviation Handbook*, figure 4-6

Creating Weight and Balance Reference Documents

After an aircraft has completed flight testing, engineers must produce the reference documents required by the pilots to carry out the weight and balance analysis prior to every flight they plan.

EXERCISE 4.2   THE PILOT'S OPERATING HANDBOOK

Objective

Design the required weight and balance pages for a Pilot's Operating Handbook.

Materials

- Reference charts on aircraft performance and limitations (Figure 4-6)
- Calculator
- Graph paper
- Computer
- Engineer's notebook

Procedure

STEP 1 Use the information provided to create a complete set of weight and balance documents that include one or more of the calculation, chart, and table methods for the aircraft described.

STEP 2 It may be helpful to complete some research by asking local pilots for the Pilot's Operating Handbook for their aircrafts. Many handbooks can also be found online.

DEPARTMENT OF TRANSPORTATION
FEDERAL AVIATION ADMINISTRATION

	2A13
	Revision 44
	PIPER
PA-28-140	PA-28-151
PA-28-150	PA-28-161
PA-28-160	PA-28-181
PA-28-180	PA-28R-201
PA-28-235	PA-28R-201T
PA-28S-160	PA-28-236
PA-28S-180	PA-28RT-201
PA-28R-180	PA-28RT-201T
PA-28R-200	PA-28-201T
	October 15, 1997

TYPE CERTIFICATE DATA SHEET NO. 2A13

This data sheet, which is a part of Type Certificate 2A13, prescribes conditions and limitations under which the product for which the type certificate was issued meets the airworthiness requirements of the Federal Aviation Regulations.

Type Certificate Holder   The New Piper Aircraft, Inc.
                                   2926 Piper Drive
                                   Vero Beach, Florida 32960

I - Model PA-28-160 (Cherokee), 4 PCLM (Normal Category), Approved October 31, 1960, for S/N 28-03; 28-1 through 28-4377; and 28-1760A.

Engine	Lycoming O-320-B2B or O-320-D2A with carburetor setting 10-3678-32
Fuel	91/96 minimum grade aviation gasoline
Engine Limits	For all operations, 2700 r.p.m. (160 hp)
Propeller and Propeller Limits	Sensenich M74DM or 74DM6 on S/N 28-1 through 28-1760, and 28-1760A. Sensenich M74DMS or 74D6S5 on S/N 28-1761 through 28-4377. Static r.p.m. at maximum permission throttle setting not over 2425 r.p.m., not under 2325 r.p.m. No additional tolerance permitted. Diameter: Not over 74", not under 72.5".
Propeller Spinner	Piper P/N 14422-00 on S/N 28-1 through 28-1760, and 28-1760A. Piper P/N 63760-04 or P/N 65805-00 on S/N 28-1761 through 28-4377. See NOTE 11.

Airspeed Limits

Never exceed	171 mph	(148 knots)	CAS
Maximum structural cruising	140 mph	(121 knots)	CAS
Maneuvering	129 mph	(112 knots)	CAS
Flaps Extended	115 mph	(100 knots)	CAS

Page No.	1	2	3	4	5	6	7	8	9	10	11	12	13	14	15	16	17	18
Revision No.	44	43	43	43	44	43	44	43	44	43	44	44	44	43	43	44	44	44
Page No.	19	20	21	22	23	24	25	26	27	28	29	30	31	32	33	34	35	36
Revision No.	43	43	44	43	44	43	44	43	44	44	44	43	44	43	44	44	43	43
Page No.	37	38	39	40	41	42												
Revision No.	43	43	43	44	44	43												

FIGURE 4-6   *Type Certification 2A13*

Source: Federal Aviation Administration

© Cengage Learning 2012

(continued)

Center of Gravity Range	(+84.0) to (+95.9) at 1650 lb. or less
	(+85.9) to (+95.9) at 1975 lb.
	(+89.2) to (+95.9) at 2200 lb.
	Straight line variation between points given.

Empty Weight C. G. Range   None

Maximum Weight   2200 lb.

No. of Seats   4   (2 at +85.5, 2 at +118.1)

Maximum Baggage   125 lb. at (+142.8) on S/N 28-1 through 28-1760, and 28-1760A.   See NOTE 8.
     200 lb. at (+142.8) on S/N 28-1761 through 28-4377.

Fuel Capacity   50 gallons at (+95)   (2 wing tanks)
     See NOTE 1 for data on system fuel.

Oil Capacity   8 quarts at (+32.5), 6 quarts usable
     See NOTE 1 for data on system oil.

Control Surface Movements

Wing flaps	(±2°)	Up	0°	Down	40°
Ailerons	(±2°)	Up	30°	Down	15°
Rudder	(±2°)	Left	27°	Right	27°
Stabilator	(±2°)	Up	18°	Down	2°
Stabilator Tab	(±1°)	Up	3°	Down	12°

Nose Wheel Travel

(±1°)	Left	30°	Right	30°

(Effective on S/N 28-1 through 28-3377, and 28-1760A)

(±1°)	Left	22°	Right	22°

(Effective S/N 28-3378 through 28-4377)

Manufacturer's Serial Nos.   28-03; 28-1 through 28-4377; and 28-1760A.

II - Model PA-28-150 (Cherokee), 4 PCLM (Normal Category), Approved June 2, 1961, for S/N 28-03; 28-1 through 28-4377; and 28-1760A.

Engine   Lycoming O-320-A2B or O-320-E2A with carburetor setting 10-3678-32

Fuel   80/87 minimum grade aviation gasoline

Engine Limits   For all operations, 2700 r.p.m.   (150 hp)

Propeller and Propeller Limits   Sensenich M74DM or 74DM6 on S/N 28-1 through 28-1760, and 28-1760A.
     Sensenich M74DMS or 74DM6S5 on S/N 28-1761 through 28-4377.
     Static r.p.m. at maximum permissible throttle setting not over 2375 r.p.m.,
        not under 2275 r.p.m.
        No additional tolerance permitted.
     Diameter: Not over 74", not under 72.5".

Propeller Spinner   Piper P/N 14422-00 on S/N 28-1 through 28-1760, and 28-1760A.
     Piper P/N 63760-04 or 65805-00 on S/N 28-1761 through 28-4377.
     See NOTE 11.

Airspeed Limits

Never exceed	171 mph	(148 knots)	CAS
Maximum structural cruising	140 mph	(121 knots)	CAS
Maneuvering	129 mph	(112 knots)	CAS
Flaps Extended	115 mph	(100 knots)	CAS

FIGURE 4-6   *(continued)*

CHAPTER 5
Propulsion

Skills List

After completing the activities in this chapter, you should be able to:

- Use appropriate tools to accomplish a mechanical task

- Disassemble, inspect, and reassemble a complex device

- Inspect components for wear, visual corrosion, and damage

- Apply the concept and mathematics of torque for analyzing forces in action

- Design, build, test, and utilize an experimental apparatus for analyzing the force of thrust

Internal Combustion Engines, Propellers, and Thrust Testing

The propulsion system for an aerial vehicle transforms one form of energy into another to produce a mechanical force that pushes the vehicle in an intended direction. To accomplish this, the system requires both an engine (or motor) and a device such as a propeller, nozzle or fan (Figure 5-1). In this chapter, you will experience a common engine for an RC aircraft by tearing it down for inspection and then rebuilding it. You will also design, build, and utilize a testing device to evaluate various propellers for their suitability on a particular aircraft.

FIGURE 5-1 *A two-stroke radio-controlled glow fuel engine with propeller mounted on Nexstar aircraft.*

Source: Photo courtesy of Ben Senson

EXERCISE 5.1   BREAK IT DOWN, BUILD IT UP—AN ENGINE DISSECTION

The engine for a radio-controlled aircraft is a relatively simple device. The model that you will work with today is a two-stroke engine meaning that combustion occurs every time the piston reaches its highest position in the cylinder. These engines have few parts (Figure 5-2). The particular engine that you are working with may vary from that of the other powerplant mechanics.

Objective

For this activity, you will completely disassemble the engine to its basic components, inspect the components, suggest repair and remediation methods, and then reassemble the engine.

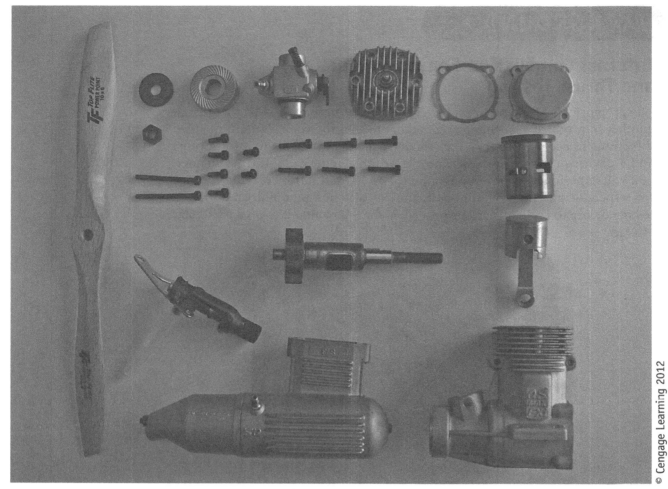

FIGURE 5-2 *The same two-stroke engine completely broken down into its individual parts.*

Source: Photo courtesy of Ben Senson

TIP SHEET

Remember to always use the correct tool for the job to be completed.

Materials

- RC glow engine
- Screwdriver, Phillips (various sizes)
- Screwdriver, regular (various sizes)
- Hex wrenches, English (various sizes)
- Hex wrenches, metric (various sizes)
- Needle-nose pliers, padded
- Pliers, padded
- Block of wood
- Rubber mallet

Procedure

STEP 1 Draw a sketch of the RC glow engine that you will be disassembling.

STEP 2 Develop a method for safeguarding and storing the engine components as they are removed from the engine. Document this method here.

STEP 3 Begin the disassembly by following these steps:

1. Remove the muffler (Figure 5-3).

2. Remove the carburetor assembly.

3. Identify the screws holding the cylinder head to the cylinder. Remove them by loosening one screw slightly, and then skip one screw before loosening the next screw until all screws are loosened. Remove all screws, and pull the cylinder head. Watch for and safeguard the gasket if one is used.

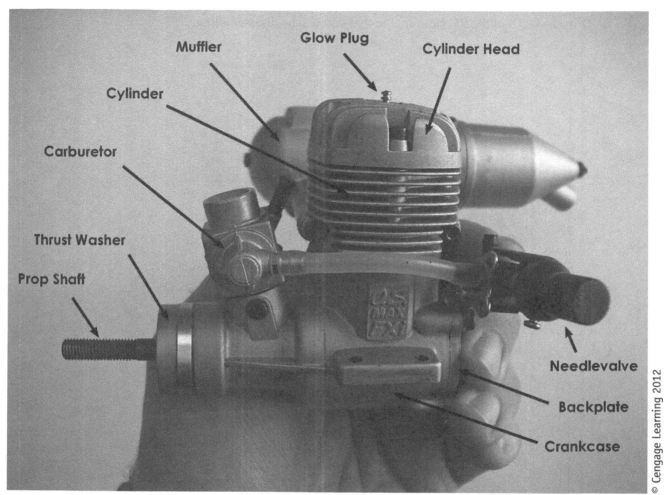

FIGURE 5-3 *Major components of a two-stroke glow fuel engine.*

Source: Photo courtesy of Ben Senson

4. Identify and remove the screws holding the engine backing plate to the crankcase. Remove the crankcase, and note its orientation carefully here. Make a short note or draw a simple diagram. Watch for and safeguard the gasket if one is used.

5. Turn over the motor's shaft by slowly turning the propeller (or prop shaft) until the piston is as low as possible in the cylinder. Insert a small, flat, firm, nonmetallic object such as the end of a popsicle stick, thin dowel, or plastic coated paperclip through the exhaust port (the one where the muffler was mounted) from the inside of the cylinder. Insert the object just far enough so

that it catches the edge of the cylinder sleeve without contacting the aluminum cylinder (Figure 5-4). Slowly turn over the motor's shaft to raise the piston and push the cylinder sleeve out of the cylinder. Carefully note and document the alignment of the sleeve in the cylinder (there are typically identifying marks, nicks, or identifiable patterns to the shape and location of the ports).

Sketch Here

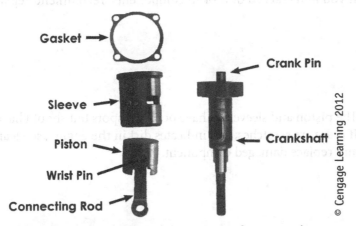

FIGURE 5-4 *Major internal components of a two-stroke glow fuel engine.*

Source: Photo courtesy of Ben Senson

6. Gently remove the cylinder sleeve. *Do not* use metallic tools directly on the sleeve. Pull gently with your fingers, or use padded tools to pull the sleeve.

7. Turn over the motor to rotate the crankshaft and the crankpin to its lowest position, pulling the connecting rod and piston downward in the cylinder. Note the orientation of the piston in the cylinder, gently pull the connecting rod off of the crankpin, and then remove the piston from the cylinder. If you need to use a tool to remove the connecting rod, use padded pliers and pull gently. Separate the connecting rod from the piston after you remove them from the engine by pushing the wrist pin out of the piston.

8. If the crankcase has two parts, separate them, and remove the crankshaft. For one-piece crankcases, remove the prop nut, washer, prop, and thrust washer (the one with the knurled surface behind the propeller), and then tap the crankshaft out of the ball bearings with a rubber mallet or with gentle pressure on the prop shaft (use a block of wood so you don't damage the table surface). Note the port in the crankshaft . . . write a brief description of its function based on its location in the engine assembly.

Explain Here

> **STEP 4** Inspect all components for corrosion, wear, and damage. In the spaces provided below, Note any issues discovered, and draft an inspection report that summarizes the damage or problem identified as well as suggested repair and remediation steps to be taken.

1. If you find cracked or broken components, recommend replacement.

2. The piston and sleeve can have polished spots but should have no scratches. If there are scratches, this indicates dirt in the engine, so clean out engine and replace damaged component.

3. If the piston has rings, they should be complete and without nicks. Replace both rings and sleeve if necessary.

4. The backplate may have shallow scratches but nothing deep. If necessary, replace bearings, crankshaft, and backplate.

5. If a castor oil "varnish" appears on components, remediation is to use a castor/synthetic blend.

6. If the shaft is loose or wiggles after assembly is complete, replace the bearings.

STEP 5 ▶ Reassemble the motor while lightly lubricating moving components with a few drops of glow fuel, castor oil, or 3 in 1 lubricant.

STEP 6 ▶ Confirm that the motor turns over smoothly after the assembly is completed.

STEP 7 ▶ Have your motor inspected by the instructor.

TIP SHEET

Flammable: Glow fuel is a mixture of the energy providing fuel for combustion as well as oil for lubricating the engine, protecting from corrosion, sealing microgaps, and preventing foaming in the engine. Formulations commonly contain 10 to 20% nitromethane, a very easily combusted substance. Use caution when using glow fuel, and avoid sources of ignition.

BACKGROUND

Pushing It to the Limits—Measuring Thrust

When designing an aircraft, aerospace engineers have to consider how the design will create sufficient lift to provide forces capable of overcoming the weight of the vehicle and for maneuvering the vehicle through space. They also have to take into account that any movement through a medium creates drag forces that tend to transform the energy of motion so that the vehicle is brought to rest. To overcome drag and sustain long duration flight, engineers use engines and motors in combination with propellers and fans to produce a forward force that overcomes drag. This forward force is called thrust.

In this activity, you will use an electrical motor and propeller "powerplant" to produce a thrust force. You will design, build, and utilize an experimental test setup to determine at least one of the following factors:

1. Thrust that is produced by propellers of differing pitch at full throttle
2. Thrust that is produced by propellers of differing diameter at full throttle
3. Energy efficiency for producing a constant amount of thrust using propellers of various pitch
4. Energy efficiency for producing a constant amount of thrust using propellers of various diameter
5. The impact of a relative wind on the amount of thrust produced by propellers of various pitch at full throttle

EXERCISE 5-2   THRUST STAND TESTING

Objective

Design an experimental device that allows you to safely and reliably measure the thrust produced by an electrically powered powerplant system consisting of a motor and propeller (Figure 5-5).

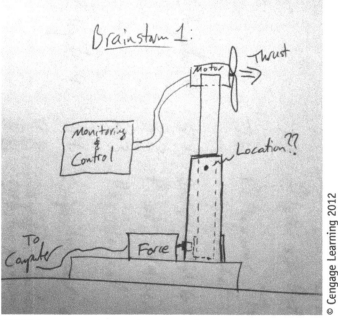

FIGURE 5-5 *Brainstorming sketch for an engine test stand without any scale or dimensions.*

Materials

- Engineer's notebook
- Graph paper
- Force measuring devices with computer interfaces
- Construction materials

Procedure

STEP 1 Determine the dimensions for all components of the powerplant provided.

STEP 2 Design a mounting system that allows you to securely fasten the powerplant to the test apparatus.

STEP 3 Determine the range and precision of the force-measuring equipment to which you have access.

TIP SHEET

Moments and Scaling Forces

A moment is the ability of a force to induce rotation around a pivot point such as the center of gravity of a free body or a physical hinge point for a mounted device. Similar to linear motion, moments only produce rotational acceleration when there is a net moment—such as when two twisting forces (moments) are being applied in opposite directions, but one of them is greater than the other, or when only one moment is being applied. If we know that moments are being applied to an object, but it is not rotating, or it is rotating at a constant speed, then the moments in one direction are equal to the moments in the opposing direction.

Calculating a Moment

The moment is calculated by multiplying the distance along the moment arm, measured from the point of force application to the point of rotation, by the force applied perpendicular to that arm. (See textbook Chapter 4 for details.)

STEP 4 Determine whether your force-measuring devices can directly connect to the powerplant system to measure thrust produced and if not, then by what ratio the forces need to be scaled in order to be measurable with this equipment.

TIP SHEET

Caution: Remember that even if our equipment is capable of being directly connected to the powerplant creating the thrust force, we will need to design a mounting system that braces the powerplant in alignment and prevents it from separating from the test stand. Always design for safety first!

STEP 5 Brainstorm a number of possible solutions to this design challenge. Sketch and dimension the devices that you propose to use for measuring the thrust of the powerplant. Incorporate these sketches and a mathematical analysis in a design brief for building your testing device. Submit the proposed solution to your instructor for approval.

STEP 6 Build your device, and run preliminary tests at low speeds to ensure that the design is safe and reliable.

STEP 7 Design the experiment to gather, analyze, and summarize the data related to your research question.

TIP SHEET

Measuring Electrical Energy

A measure of efficiency is the amount of energy required to carry out an identical task or the rate at which energy is consumed to perform the task. For this research project, this may mean that you determine the amount of electrical energy, or the power, per unit of thrust produced. To determine the rate of electrical energy being used by the powerplant we need to determine two things: voltage and current. Voltage is measured with a multimeter wired in parallel to the powerplant while current is measured in series with the powerplant circuit. The power, or amount of energy per second, consumed by the circuit is then just the product of these two values. If our measurements are in volts and amps, then the power will be in watts.

STEP 8 Run the experiment.

STEP 9 Write a technical report summarizing your findings.

STEP 10 Gather with the other research teams working on the same key question. Work to produce a presentation summarizing your collective findings. Present to the entire class.

CHAPTER 6
Avionics and Other Flight Systems

Skills List

After completing the activities in this chapter, you should be able to:

- Diagram and explain the operation of an Instrument Landing System (ILS)

- Diagram and explain the operation of a very high frequency omnidirectional range system (VOR)

- Demonstrate proficiency in the use of radio navigation instrumentation to navigate an aircraft to a successful landing in simulated Instrument Flight Rules (IFR) conditions

- Describe the training and certification requirements for earning the various required skills to become an Airline Transport Pilot (ATP)

Flying in Instrument Conditions

The instrumentation in the cockpit of a modern airplane is designed to provide the pilot with information about its orientation and movement in space as well as the operation of its systems. Knowing where the aircraft is located and how it is moving through space is referred to as spatial orientation. To maintain spatial orientation, pilots invest a significant effort in training and practicing to make proper use of navigational instruments. These skills are critical for allowing pilots to continue a flight beyond visual into Instrument Flight Rule (IFR) conditions. During IFR flights, the navigational instrumentation is capable of providing all the information necessary to maintain spatial awareness. In this chapter, you will become familiar with both VOR and ILS aids to navigation and how they are used to maneuver an aircraft during cruising flight and for the approach to landing at an airport (Figure 6-1). You will also become familiar with the training experiences required to become a certified Airline Transport Pilot (ATP).

© Cengage Learning 2012

FIGURE 6-1 *Airliner using the ILS localizer to stay lined up for final approach.*

A very high frequency omnibearing radio range station provides the means to navigate by creating a combination of radio signals to identify the direction along a radial on which the aircraft is currently located and how to navigate to intercept any other radial from that station. The system works by broadcasting an FM omnidirectional signal in all directions every time a rotating AM beacon moves through magnetic north. If a receiver picks up both signals aligned perfectly in time (in phase), then the instrument knows that you are currently on the radial pointing magnetic north of the station. Anywhere else around the station, the timing of the AM signal is delayed (out of phase). This delay is by exactly the same number of degrees as your magnetic azimuth from the station (Figure 6-2). Using two stations simultaneously allows a pilot to pinpoint his or her location to within a 1-degree accuracy.

V.O.R. Phase Shift

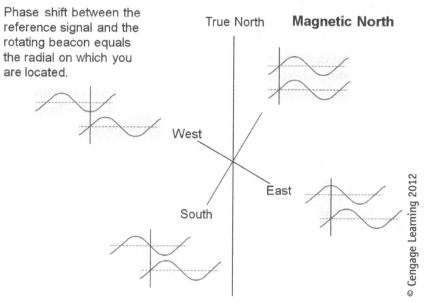

Phase shift between the reference signal and the rotating beacon equals the radial on which you are located.

FIGURE 6-2 *VOR systems broadcast both a reference omnidirectional signal and a rotating signal. The difference in phase matches that of the magnetic radial on which the receiver is located.*

EXERCISE 6.1 NAVIGATING BY VOR

Objective

Become skilled in the use of VOR equipment for navigating an aircraft.

Materials

- Internet access
- Flight Simulator
- "Tim's Air Navigation Simulator" Java applet (download available at www.visi.com/~mim/nav/)
- Local sectional chart for planning navigation (http://skyvector.com/)

Procedure

STEP 1 Research the components of a VOR navigation system, and sketch and explain the function of each component in your engineer's journal. Include the following:

- VOR station
- Antenna
- NavCom Radio
- Omnibearing Selector (OBS)
- Course Indicator
- Course Deviation Indicator
- To/From Indicator

TIP SHEET

Microsoft Flight Simulator is a tremendous resource. In the Learning Center, maneuver to the Navigation section, and explore the information it contains on pilotage, dead reckoning, VORs, GPS, and much more.

STEP 2 Identify a local VOR station, and find its frequency and Morse code identifier.

STEP 3 In Flight Simulator, place your aircraft near this VOR station at an altitude of at least 1000 ft above ground level (AGL) and with sufficient airspeed to maintain altitude. Start the flight, and tune your Nav 1 radio to the VOR. Listen for its Morse code identifier to confirm you are tuned into the correct station. If you do not hear the identifier, confirm the frequency is correct, has been made active, and you are sending the Nav 1 radio to the speakers. If there isn't an identifier, the station is out for maintenance. This happens in the real world but not in Flight Simulator.

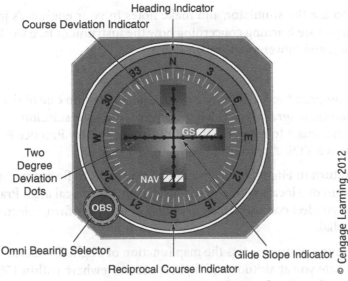

Heading Indicator

Course Deviation Indicator

Two Degree Deviation Dots

GS

NAV

OBS

Omni Bearing Selector

Reciprocal Course Indicator

Glide Slope Indicator

© Cengage Learning 2012

FIGURE 6-3 *Components of the ILS instrument face.*

STEP 4 To gain experience with the VOR instrumentation and its relationship to the VOR station and its radials, Google to find "Tim's Air Navigation Simulator" and run the Java applet. Note that the instrument panel on the right has two VOR instruments (Figure 6-3) that are color coded to indicate that the upper instrument has been tuned to the frequency for the red station while the lower instrument is tuned to the blue station.

Use the arrow keys in the lower corners of the red instrument to rotate its omnibearing selector (OBS) knob and rotate the course indicator ring until the course deviation needle is centered. Note that there are two headings at which the needle is centered: with one, the TO indicator is illuminated, and for the other the FROM indicator is illuminated.

The VOR only tunes one radial at a time. When the needle is centered with the FROM indicator illuminated, the radial tuned runs right through the aircraft. If you turn to the heading that matches the radial that is tuned, the needle would stay centered as you fly outward from the station. Use the simulator to set up situations like this. Note that you can toggle the wind mode on and off by pressing "W."

If you drift off of the radial, the course deviation needle deflects one dot for every two degrees of angle you are off the radial and in the direction you need to change your heading in order to get back onto the radial.

TIP SHEET

Danger! The VOR has no way of knowing the current heading or orientation of the aircraft. It deflects the course deviation needle with the assumption that you have rotated the omnibearing selector to place your intended course at the top of the instrument. Use caution and confirm that you always tune the VOR to the radial you intend to fly away from the station, and set your heading to match before visualizing which way to turn to intercept the radial.

Continue to use the simulator, and make notes in your engineer's journal about the lessons you are learning concerning how the instrument face works to indicate your location and movement through space.

STEP 5 ▶ Complete the VOR Practice Worksheet based on one of the four VOR Station diagrams provided. After verifying your solutions, partner with a classmate to complete the VOR Cross Country Practice Problem or plan a VOR Approach to Landing.

STEP 6 ▶ Return to Flight Simulator, and use the VOR to fly a straight line course from one location to another location in your local area. Practice until you feel confident in your ability to tune, confirm, select, and fly a radial.

STEP 7 ▶ Have a partner access the map function of Flight Simulator to randomly locate you at altitude and flight speed somewhere within 150 miles of your home town. Tune each VOR in the aircraft to a different VOR in the area, and rotate the omnibearing selectors until both needles are centered with FROM illuminated. Plot out your location on a sectional chart for your area. Pause the simulation to plan how to use the VORs to maneuver to your home airport, then unpause the simulation, and execute your plan.

TIP SHEET

Note that VOR stations are frequently used as the navigational fix from which you begin your approach to an airport to start an Instrument Landing System (ILS) precision approach to a runway.

VOR Practice Worksheet

Directions: For each aircraft location refer to the VOR Station diagram in order to accurately diagram the heading indicator, to/from indicator, and course deviation indicator onto the instrument face and estimated heading for radial intercept (assume a 45° angle).

Aircraft Location	VOR Settings	VOR Indications	Summary
A	**Tuned to VOR Station Morse code confirmed Intended Course = 150°**		**Reciprical Course = ____°** **Heading for Intercept = ____°**
B	**Intended Course = 360°**		**Reciprical Course = ____°** **Heading for Intercept = ____°**
C	**Intended Course = 220°**		**Reciprical Course = ____°** **Heading for Intercept = ____°**
D	**Intended Course = 40°**		**Reciprical Course = ____°** **Heading for Intercept = ____°**

© Cengage Learning 2012

VOR Station 1

VOR Station 2

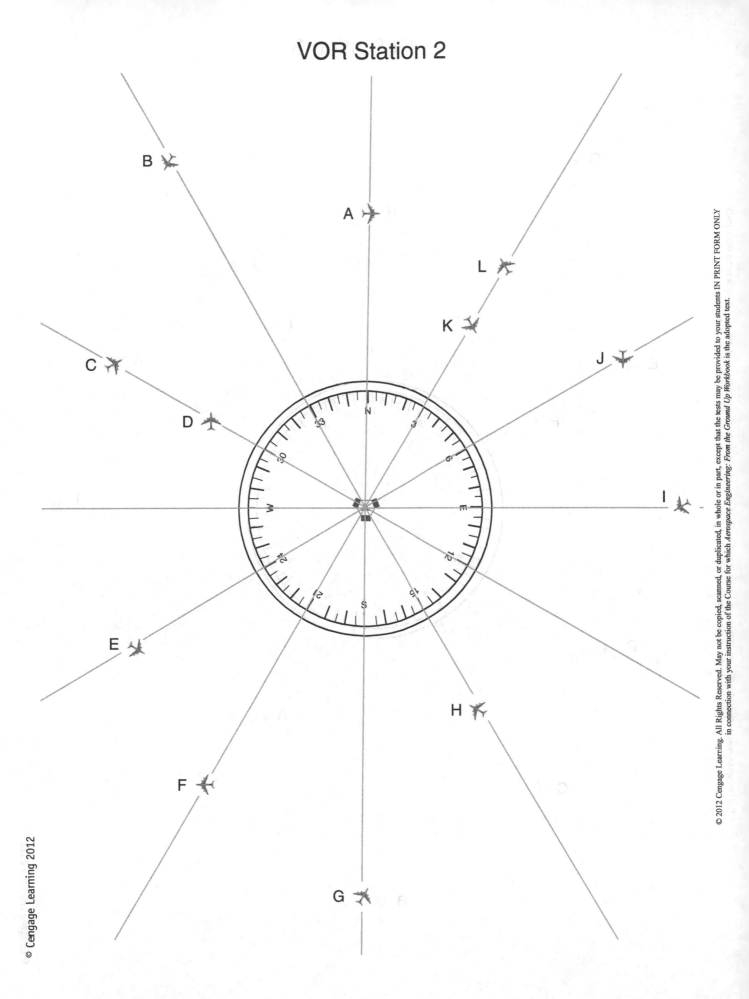

VOR Station 3

G

H

J

I

E

K

F

N

33   3

30   6

W   E

24   12

L

21   15

D   S

A

C

B

VOR Station 4

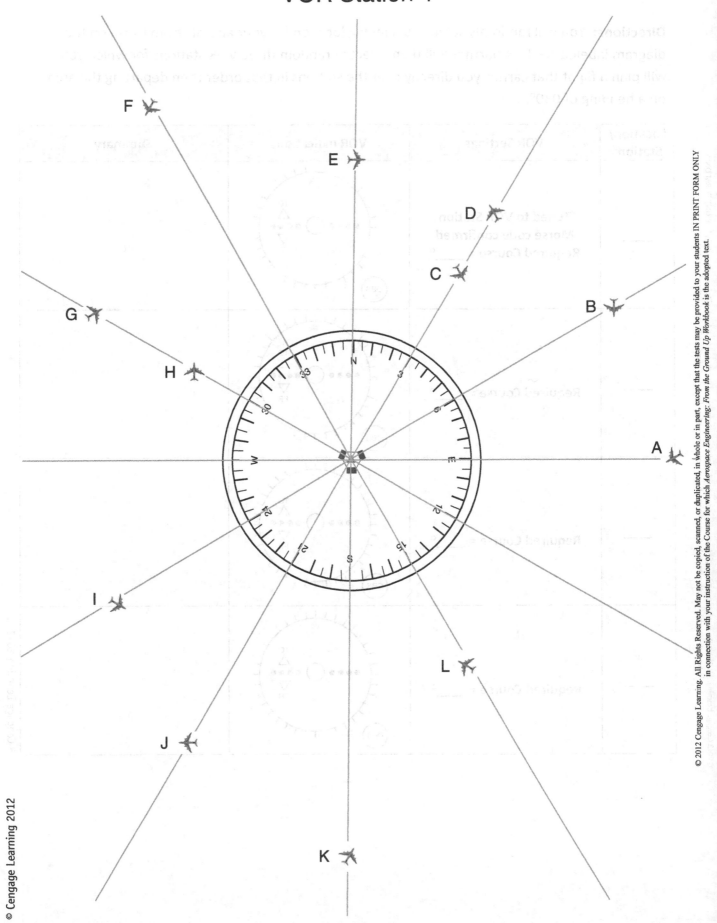

VOR Cross Country Planning

Directions: You will randomly select a beginning location for your aircraft from those on the diagram labeled A – L. A partner will then select at random three VOR stations for which you will plan a flight that carries you directly over the stations in that order then departing the area on a heading of 030°.

Location/ Stations	VOR Settings	VOR Indications	Summary
____	**Tuned to VOR Station Morse code confirmed Required Course = ____°**		
____	**Required Course = ____°**		
____	**Required Course = ____°**		
____	**Required Course = ____°**		

VOR Cross Country 1

VOR Approach to Landing 1

Instrument Landing Systems

The ILS provides accurate guidance to landing for one runway. The guidance information it provides is accurate enough for both aligning with the centerline of the runway (localizer) and following the descent path of the glide slope to the touchdown zone of the runway (Figure 6-4). In this exercise, you will complete research on how an ILS system is designed.

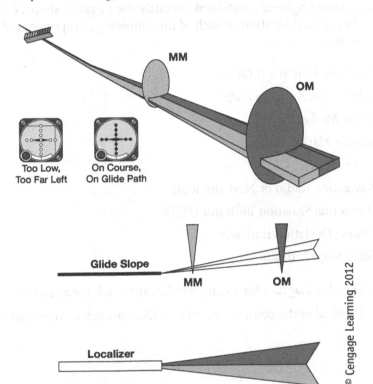

© Cengage Learning 2012

FIGURE 6-4 *Layout of a typical Instrument Landing System (ILS) for a runway.*

EXERCISE 6.2   IDENTIFY THE COMPONENTS AND FUNCTIONS OF AN ILS APPROACH SYSTEM

Objective

Gain an understanding of the ILS approach to a runway by identifying all of the components of the system and their functions.

Materials

- Internet search tools
- Aviation reference books
- Poster board or plotter
- Glue

TIP SHEET

Confirm your information! Trusting the Internet is a challenge. Talk with your librarian about methods for finding reliable information and confirming sources.

Procedure

STEP 1 In your engineer's notebook, identify the physical shape, operation, theory, and location of each of the following components of an ILS system.

- Localizer Antenna (LOC)
- Glide Slope Antenna (G/S)
- Outer Marker
- Middle Marker
- Inner Marker
- Navigation Radio or NavCom Radio
- Horizontal Situation Indicator (HSI)
- Course Deviation Indicator
- Glide Slope Indicator

STEP 2 Sketch a diagram for a complete ILS approach for a runway.

STEP 3 Label all of the components of the ILS approach in your sketch.

Designing an Approach System

In this exercise, you will interpret the approach plate for an ILS approach for a local or regional runway. You will make a presentation to the airport control board with your proposal for the ILS system design.

EXERCISE 6.3   APPROACH PLATES

Objective

Become familiar with the ILS approach plates for existing runways in your area.

Materials

- Internet access
- Printer
- Beginner's Guide to the Instrument Landing System (ILS) Approach Plate (Figure 6-5)

The Beginner's Guide to the Instrument Landing System (ILS) Approach Plate

Exemplar approach plate can be downloaded at http://www.airnav.com/depart?http://204.108.4.16/d-tpp/1101/00245ILD36.PDF

HEADER: This section includes three rows of information

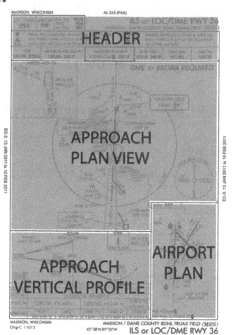

Row 1	MADISON, WISCONSIN AL-245 (FAA)
	LOC I-MSN 109.9 / APP CRS 002° / Rwy Idg 8005 / TDZE 862 / Apt Elev 887 — ILS or LOC/DME RWY 36 / MADISON / DANE COUNTY RGNL-TRUAX FIELD (MSN)
Row 2	When ALS inoperative, increase Cat E S-ILS ¼ mile and S-LOC ½ mile. • Vis Cat A/B/C/D RVR 1800 authorized with the use of FD or AP or HUD to DA. ASR — MALSR — MISSED APPROACH: Climb to 2700 via MSN VORTAC R-359 to DECAL Int/MSN 14.3 DME and hold.
Row 3	ATIS 124.65 278.3 / MADISON APP CON * 135.45 343.7 / MADISON TOWER * 119.3 (CTAF) 257.8 / GND CON 121.9 348.6 / CLNC DEL 121.62 / UNICOM 122.95

Row 1: Runway Information

LOC I-MSN **108.9**	Frequency of the Localizer
APP CRS **002**	Approach Course to the Airport
8005	Runway Landing Length
862	Touch Down Zone Elevation
887	Airport Elevation
ILS or LOC/DME	Procedure Type
RWY 36	Runway Name
MADISON / DANE COUNTY RGNL-TRUAX FIELD (MSN)	Airport Name

Row 2: Approach Information

Additional Information that describes variables for this approach

MALSR — Medium intensity Approach Lighting System with Runway Indicator lights

Written "MISSED APPROACH" procedure for this approach path.

Row 3: Frequencies for Communications

ATIS	Automated Terminal Information Services (for weather and conditions)
APP CON	Approach Controller
TOWER	Tower Controller
GND CON	Ground Controller
CLNC DEL	Clearance Delivery (for filing, opening, closing, modifying flight plan)

Approach Plan View: Overhead perspective

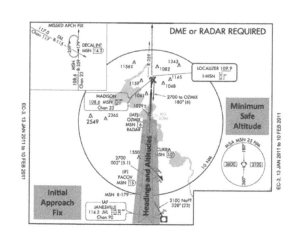

IAF: Initial Approach Fix... START HERE!

JANESVILLE VOR with a minimum descent altitude of 3100 feet until you capture the localizer and descend to 2700 feet.

Headings and Minimum Altitudes:

The approach path is on a heading of 002 degrees.
Approaching CUKRA maintaining at least 2700 feet
Continue to OZMIX maintaining at least 2500 feet
Intercept the glide slope at OZMIX

Minimum Safe Altitudes within 25 NM of MSN VOR

In order to be at least 1000 feet above any obstruction...
West of the VOR maintain 3600 feet
East of the VOR maintain 3100 feet

Approach Vertical Profile:

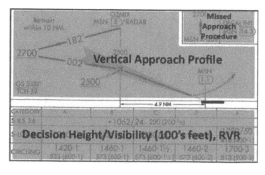

Decision Height and Visibility:

The bottom section lists the elevation which the pilot in command must maintain until they declare a missed approach or commit to the landing.
The second number is required visibility in 100's of feet
The third number is "Runway Visibility Range" required

Missed Approach Procedure:

Step by step action blocks detail the missed approach process

FIGURE 6-5 *Cheat sheet for interpreting ILS approach plates*
Source: author Ben Senson

Procedure

STEP 1 ▸ Find the airport description for your local or regional airport. You can find your local airport by searching at www.airnav.com/airports/.

STEP 2 ▸ Scroll down to the Instrument Procedures section of the airport description to find a list of all of the approaches available for pilots operating under Instrument Flight Conditions.

STEP 3 ▸ Find and download a runway with an ILS approach, and either save or print (check with your instructor) the approach plate.

STEP 4 ▸ Using the tutorial guide sheet provided, find all of the information required for executing an ILS approach to this runway (Figure 6-6).

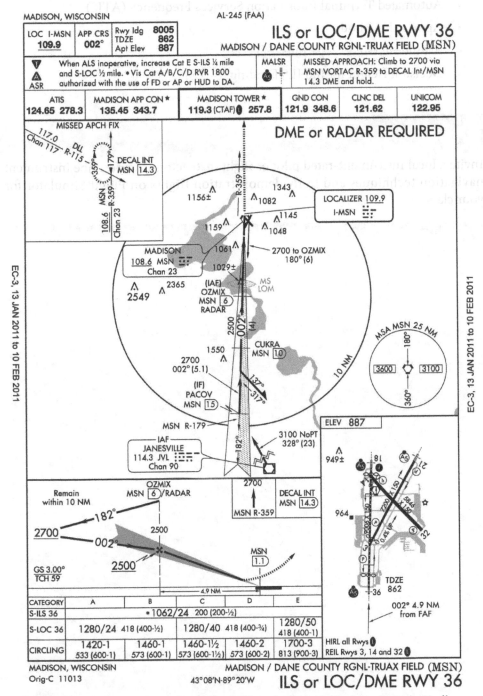

FIGURE 6-6 *Example of an ILS approach plate for one particular runway in Madison, Wisconsin.*

Source: FAA (Federal Aviation Administration)

Approach Information:

- Airport Location
- ILS Frequency (I-DSZ)
- Approach Course (APP CRS)
- Runway Length (Rwy ldg)
- Touch Down Zone Elevation (TDZE)
- Airport Elevation (Apt Elev)
- Runway Approach Type
- Missed Approach Procedure
- Automated Terminal Information Services Frequency (ATIS)
- Approach Altitude
- Altitude at Glide Slope Intercept
- Decision Height for full ILS Landing (S-ILS)

TIP SHEET

Invite a local instrument-rated pilot or flight instructor in to describe instrument navigation techniques and to fly demonstration flights on Flight Simulator for your class.

Flying an Actual (Simulated) Instrument Landing System Approach

In this exercise, you will make use of flight-simulation software to practice an ILS approach to landing under ideal and weather conditions that are at "minimums" allowed for a safe landing.

EXERCISE 6.4   FLYING THE APPROACH

Objective

Fly the ILS approach to a local runway.

Materials

- Flight Simulator software
- Local sectional chart
- Approach plate printout from Exercise 6.3

Procedure

STEP 1 Open flight-simulation software, and select a common general aviation aircraft such as a variant of the Cessna 172.

STEP 2 Adjust conditions to be noon local time, clear skies, and calm with no winds.

STEP 3 Open the map feature, and locate your aircraft so that it is airborne, at flight speed, and located for an extended straight-in approach to the runway you have selected. Place it far enough away that you have time to stabilize the aircraft in altitude and heading (15-20 NM).

STEP 4 For the duration of the approach to decision height (DH), try to look only at the instrument panel of the aircraft.

STEP 5 Tune the NAV 1 radio to the ILS frequency, and switch it to active.

STEP 6 Establish and maintain the approach altitude and course heading until the localizer and course deviation needle become active.

TIP SHEET

You can use the ident feature of the radio to listen to the Morse code signal of the localizer to confirm that the navigation radio is properly tuned.

STEP 7 Continue your approach to the glide slope intercept and descend along it toward the DH indicated on the ILS approach plate for this runway.

STEP 8 As you approach the DH, look out the window and confirm that you can visually see the runway or its approach lighting.

STEP 9 Make the decision to commit to the landing and land the aircraft visually or "declare a missed approach" and increase power, raise flaps, and reset the Flight Simulator for another attempt.

STEP 10 After you have mastered the basics of the ILS approach for this runway, fly a more realistic IFR approach by returning to the settings for the Flight Simulator and changing the conditions to midnight with 100% overcast, cloud base at the DH for the approach, and visibility to match that of the minimums indicated on the approach plate. Add in gusty winds and light turbulence if you are feeling brave.

STEP 11 The runway you prepared for has been closed for repainting and you have to change at the last second to an alternative runway. Have a classmate select a different ILS approach to the airport. Set up the aircraft's location for approximately 15 NM out with altitude, airspeed, and heading for an extended approach to the alternate runway. Start the flight, and then reference the new approach plate while flying the airplane. Continue the flight without pausing the simulation as you locate the correct approach plate and execute the approach and landing on the new runway.

BACKGROUND

Earning Your Wings

The training and experiences required to earn and maintain a pilot's license vary with the level of skill required to safely operate the aircraft under typical and emergency flight conditions. Pilot's progress from flying for pleasure or personal business to commercial flight and can eventually become sufficiently skilled to earn the right to fly large numbers of passengers as Airline Transport Pilots (ATP) (Figure 6-7). In this exercise, you will investigate and summarize the licensure requirements to transform a beginning pilot into an ATP.

© Cengage Learning 2012

FIGURE 6-7 *A pilot's wings.*

EXERCISE 6.5   BECOMING AN AIRLINE TRANSPORT PILOT

Objective

Become familiar with the career requirements for becoming "airline ready" and ATP certified.

Materials

- Internet access
- Local flight training school or certified flight instructor (CFI)

Procedure

STEP 1 Research and summarize the required training, knowledge tests, practical tests, and hours of training required to earn the following pilot certification levels:

- Student Pilot
- Sport Pilot
- Recreational Pilot
- Private Pilot
- Certified Flight Instructor
- Instrument Rated Pilot
- Certified Flight Instructor Instrument
- Commercial Pilot
- Multi-Engine Pilot
- Airline Transport Pilot

STEP 2 ▶ Describe the requirements for earning and maintaining the medical certifications and match them to their respective pilot certification level.

- Third Class Medical
- Second Class Medical
- First Class Medical

STEP 3 ▶ Optional: Earn your student through Airline Transport Pilot licenses through the Microsoft Flight Simulator Flight Lessons.

CHAPTER 7
Astronautics

Skills List

After completing the activities in this chapter, you should be able to:

- Use test equipment to measure the thrust force produced by a rocket engine

- Test a representative sample of a product to determine its variability and reliability in meeting specified performance characteristics

- Integrate data visualization and video materials into a single test summary product

- Design rocket components and integrate them into a complete rocket

- Evaluate a rocket design for stability in flight

- Predict the flight-performance characteristics for a custom rocket

- Build and quality inspect a model rocket

- Understand and comply with safe launch practices for model rockets

BACKGROUND

Rocket Design from Thrust Test to Launch

Rockets are the standard propulsion system for missiles and spacecraft. Small, reliable engines are commercially available for use in model rockets (Figure 7-1). In this chapter, you will thrust test commercially available model rocket engines, produce a quality assurance report, design a custom model rocket, and evaluate the rocket's stability and predicted flight performance.

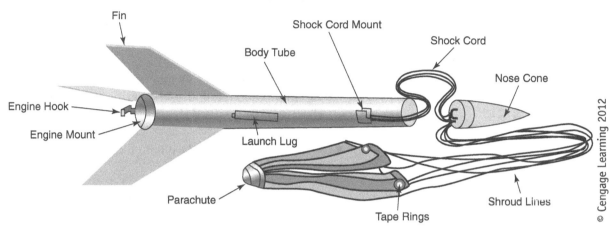

FIGURE 7-1 *Components of a typical model rocket.*

Determining the accuracy and reliability with which model rocket engines are manufactured requires that we collect data on a number of engines with the same performance specifications. Model rocket engines are solid fueled and have no moving parts (Figure 7-2). To collect performance data, we need a mechanism for mounting the motors, a device for measuring the thrust force the motor produces, and visualization and analysis software for processing the data collected.

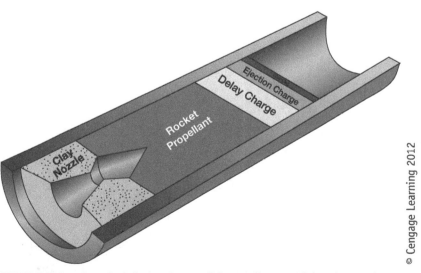

FIGURE 7-2 *A typical design for a solid propellant model rocket engine.*

EXERCISE 7.1   THRUST TESTING ROCKET ENGINES

Objective

Test commercial off-the-shelf rocket engines for the accuracy and precision of their manufacturing.

Materials

- PASCO Data Studio Software (or equivalent such as Vernier's Logger Pro software and hardware)
- PASCO Rocket Engine Test Bracket ME-6617
- PASCO PASPORT Force Sensor PS-2104
- PASCO Datalogger PS-2000 or
- PASCO Airlink SI PS-2005A
- Ring stand
- Still camera on tripod (.bmp, .jpg, .jpeg, or .gif at 240 × 320 or 480 × 640)
- Video camera (.mov, .avi, .mpeg, .mpg recorded at 240 × 320 or 480 × 640)
- Leather gloves
- Estes model rocket engines (12 each of):
  - A8-3
  - B6-4
  - C6-5
  - D12-3
- Engine igniters
- Launch controller
- Estes Engine Chart (http://www.estesrockets.com/pdf/Estes_Engine_Chart.pdf)
- Safety equipment (such as fire extinguisher, water, etc.)

TIP SHEET

Danger! Rocket engine exhaust is extremely hot, and the engines burn until all of their fuel is consumed. This makes it very important that you follow all of the safety recommendations for the use of model rocket engines as published by the National Association of Rocketry (www.nar.org/NARmrsc.html). Because your engines will be attached to a thrust stand, you must also ensure that the stand is braced or weighted down and cannot move.

Procedure

Data Collection

STEP 1 ▶ Set up the equipment as shown in Figure 7-3. Be certain to set up outside, far from any flammable materials. Have all required safety equipment at the testing site.

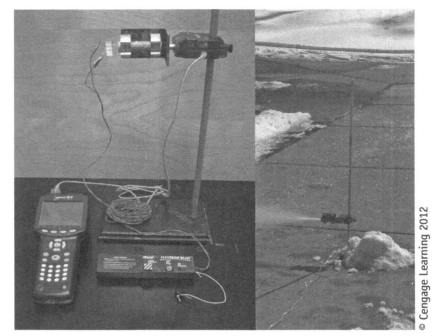

FIGURE 7-3 *An engine test stand includes equipment to firmly mount the rocket engine and record the thrust forces it creates.*

Source: Photos courtesy of Ben Senson

STEP 2 ▷ Prepare the data-collecting equipment:

1. Select the data interface: PasPort Com #.

2. Set sample rate to 250 Hz.

3. Force direction is positive.

4. Sampling options: delayed start with 1.0 N trigger and 1.0 seconds of prior data retained.

5. Sampling options: auto stop after 10 seconds.

6. Press the "zero" button on the force probe.

STEP 3 ▷ Prepare an A8-3 engine for testing by inserting an igniter and plug into the engine nozzle.

STEP 4 ▷ Insert the engine into the testing bracket.

STEP 5 ▷ Remove the safety key from the launch controller, and then connect the electrical clips to the igniter.

STEP 6 ▷ Clear the test area, insert launch key, countdown, start camera on 3, and then press and hold the safety key and ignition button for two seconds on zero.

TIP SHEET

Wait 60! If the engine doesn't ignite, pull the safety key out, and wait a full 60 seconds before approaching the launch stand to check launcher clips or to replace the igniter.

STEP 7 During the engine burn, take a still picture.

STEP 8 After the engine burns, coasts, and fires the ejection charge, wait until the full 10 seconds from ignition have gone by before stopping the video recording and approaching the test stand.

STEP 9 If necessary, save the video file and the data file prior to continuing. Maintain a log of the test run number, engine tested, and anything note-worthy about the run in your engineer's journal.

STEP 10 Repeat steps 3–9 for the B6-4, then the C6-5, and finally the D12-3 engines.

STEP 11 Collect all spent engines, igniters, and engine plugs, and store them in a metal container until disposed of in an outdoor trash bin.

Data Processing, Visualization, and Analysis

STEP 1 Open the data file in DataStudio.

STEP 2 Under the Data pull-down, unselect all but one test run, and note which engine this represents.

STEP 3 Use the cursor to drag a box over the data points from ignition to the end of the thrust burn to select them for analysis.

STEP 4 Under the Summation tool, select Area (Figure 7-4).

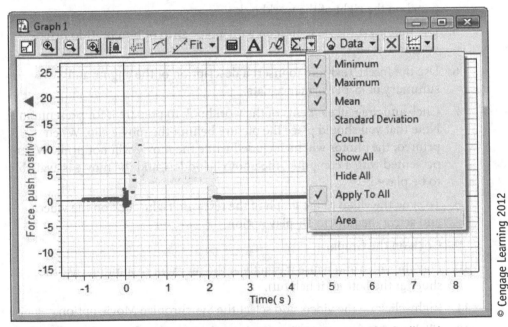

FIGURE 7-4 *Determining the total impulse of a rocket motor requires finding the area under the Force versus Time graph from the start to the end of the thrust burn.*

Source: Screen capture from DataStudio by Ben Senson

© Cengage Learning 2012

STEP 5 ▸ Note the mean thrust produced by the engine and the area under the data plot. Note that the units of area are N-s or force times distance—this is the total impulse of the engine.

STEP 6 ▸ Record the average thrust, peak thrust, duration of thrust, duration of the delay burn, and the total impulse produced by this engine.

STEP 7 ▸ Repeat for all four engines tested by creating a new graph for each engine.

1. In the Displays panel, double-click on the Graph item.
2. Choose the data source for Run #2.
3. Click OK.
4. Repeat for each data run.

STEP 8 ▸ Use Excel or similar software to organize and summarize the data collected by all groups. Include a separate labeled row or column for the published performance specifications for each engine type as found on the Estes Engine Chart.

STEP 9 ▸ Prepare a Data Studio workbook for each engine to summarize your findings.

1. In the Displays panel, double-click on the Workbook item.
2. Maximize the size of the Workbook 1 window.
3. In the Displays panel, click and drag Graph 1 into the workbook.
4. Use the Insert Picture tool (icon with blue, green, and red dot in it) to insert the picture of this test run. Press CTRL + T if you can't see the tool panel on the right of the workbook.
5. Use the Insert Movie tool (icon with a blue strip of film in it) to insert your movie clip.
6. Use the Insert Text tool to insert a description of the engine and a summary of its performance data.
7. Click and drag all elements of the workbook to prepare it for presentation. Note that you should place the picture behind the movie clip. When printed, the photos will be included while the movie will not print. When presented with a computer, the photo is hidden, and the movie is available to be played.
8. To sync the video to the graph of the test run, right-click on the video, and select the Link to Display option.
9. Click on the Graph.
10. Play the video and pause it exactly at the moment of ignition (use the slider at the bottom if helpful).
11. Right-click on the video, and select the Synchronize Movie option.
12. Click on OK to use 0.0000 seconds on the graph as the synchronization point.

STEP 10 ▸ Save all your work, and print out a copy of each workbook.

BACKGROUND

Designing for Performance

Designing an interesting shape for a model rocket is easy because it's very *subjective* and based on personal opinion. On the other hand, designing a rocket that will perform well and remain stable in flight is very *objective* because all objects have to obey the laws of physics. In this exercise, you will either design, evaluate, and then build a custom rocket from scratch, or you will take a commercially manufactured model rocket kit and evaluate and build it from the kit (Figure 7-5). In either case, you will become familiar with the characteristics of a rocket and the requirements for stability in flight.

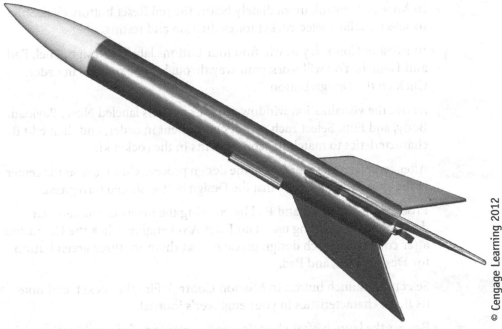

FIGURE 7-5 *A typical first model rocket project is exemplified by the Estes Alpha Kit.*

Source: Photo by Ben Senson

EXERCISE 7.2   DESIGNING A MODEL ROCKET

Objective

Predict the flight performance of a rocket.

Materials

- RocketModeler III a free Java applet (www.grc.nasa.gov/WWW/K-12/rocket/rktsim.html)
- Optional: RockSim by Apogee (www.apogeerockets.com/rocksim.asp)
- Model rocket kit such as the Estes skill level 1 "Alpha"
- Calipers
- Rulers
- Protractors
- Optional: Miscellaneous rocket components
- Optional: 3D printer

Procedure

STEP 1 ▸ Obtain a rocket kit.

STEP 2 ▸ Identify and catalog all of the major components of the model rocket kit. Obtain replacements for any missing parts.

STEP 3 ▸ Sketch, measure, and record the dimensions of each rocket component in your engineer's journal.

STEP 4 ▸ Start Rocket Modeler III.

- The left side of the applet is a visualization window.
- The upper-right part of the applet window is Mission Control.
- The lower-right part of the applet window varies by the task being accomplished.

STEP 5 ▸ In Mission Control, immediately below the red Reset button, click Solid to select a solid-fueled rocket for evaluation and testing.

STEP 6 ▸ In Mission Control, you will find four buttons labeled Design, Fuel, Pad, and Launch. You will work your way through these options in order. Click on the Design button.

STEP 7 ▸ Above the visualization window are four options labeled Nose, Payload, Body, and Fins. Select each rocket component in order, and then edit its characteristics to match the components in the rocket kit.

STEP 8 ▸ After completing all four tabs in the design process, click Go near the center of the applet window. Note that the Design button should turn green.

STEP 9 ▸ Proceed through Fuel and Pad by entering the information necessary to describe the kit making use of an Estes A8-3 engine. Click the Go button after completing each design phase so that there are three green buttons for Design, Fuel, and Pad.

STEP 10 ▸ Select the Launch button in Mission Control. Fire the rocket, and note its flight characteristics in your engineer's journal.

STEP 11 ▸ Repeat the launch after changing pad conditions for combinations of 0, 5, 10, and 15 feet per second winds and launch angles of 0, 5, 10, 15, 20, 25, and 30° into the wind. Be certain that Weathercock is turned ON. Record the maximum altitude, time aloft, and distance from the launch pad on landing for every combination of conditions.

STEP 12 ▸ Repeat steps 9–11 for both the B6-4 and the C6-5 engines.

STEP 13 ▸ For every engine used, record the launch angle that produced the closest landing to the launch pad for each wind speed. Based on the pattern of the landing spots, predict the angle of launch that should land the rocket back at the launch pad for each wind speed.

OPTIONAL: Design a custom rocket following the guidelines of the National Association of Rocketry. Use RocketModeler or RockSim software to model the behavior of your creation to prove that the design is flightworthy and to determine which engines are appropriate for use in the model. You are restricted to using unmodified commercial off the shelf (COTS) engines no larger than D12-5 in size. Check local restrictions for more restrictive limitations. Write up a design brief summarizing your design, engines to be used, predicted performance with these engines, and formal request for permission to build the rocket. If available, use Inventor and 3D printing to create custom nose cones, fin sets, body tube elements, and motor mounts.

BACKGROUND

Build and Fly a Model Rocket

We learn to trust our design process when the products that it produces work when prototyped and tested in the real world. For this exercise, you will construct the rocket that you evaluated in Exercise 7.2, collect data on its flight performance, and then compare the predicted to the actual flight performance.

EXERCISE 7.3   BUILDING A MODEL ROCKET

Objective

Build a prototype rocket and collect flight test data.

Materials

- Rocket components for the rocket evaluated in Exercise 7.2
- Glue (white glue and Modeler's wood glue or Cyanoacryolates)
- Sandpaper (medium, fine, and extra fine grit)
- Hobby knife
- Optional: Paint
- Launch pad
- Launch controller
- Igniters
- Safety equipment (fire extinguisher, water, etc.)
- Rocket engines (A8-3, B6-4, C6-5, others)
- Altitrak device or protractors with plumb bobs
- Long tape measure
- Stopwatch
- Optional: GPS device
- Optional: Still and video cameras (record launches and sync video to workbooks in Exercise 7.1)

Procedure

STEP 1 ▶ Build the rocket according to the design, plan, and instructions. Use standard construction techniques.

STEP 2 ▶ Have your rocket inspected by your instructor for quality and safety.

STEP 3 ▶ Launch rockets and collect data on flight performance. Follow all safety procedures as outlined in Exercise 7.1 and used by the National Association of Rocketry.

STEP 4 ▶ Record data required to find actual maximum altitude, distance to launch pad on landing, and time aloft.

STEP 5 ▶ Write a technical report summarizing and comparing the predicted to actual flight performance of your model rocket. Identify sources of error, and make recommendations for improving the procedures for the manufacturing and flight testing of rockets.

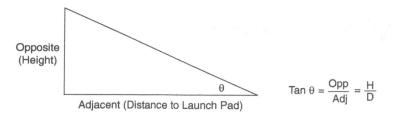

$$\text{Tan } \theta = \frac{\text{Opp}}{\text{Adj}} = \frac{H}{D}$$

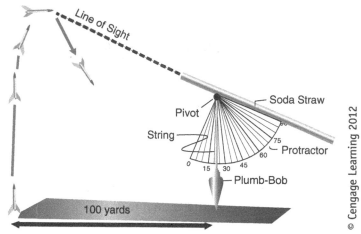

FIGURE 7-6 *Determining the height of a rocket's flight can be accomplished with a homemade angle-measuring device and by applying some simple trigonometric analysis.*

TIP SHEET

Remember Pythagoras! To measure the altitude the rocket obtained, you can use commercially manufactured devices such as the Altitrak, or you can simply measure the angle above the horizontal from a known distance away from the location of maximum height (Figure 7-6). Because it is difficult to know in advance where this point will occur, it is best practice to send at least two individuals far away from the launch pad at 90° angles to the direction from which the wind is coming and somewhat up range from the launch site. Have each observer and their assistant record the angle data for every launch observed in order as someone at the launch site records the names of each person launching their rocket in order. This makes it quick work to later calculate the maximum altitudes for each rocket and engine combination.

For measuring distance to landing sites, you can also use a GPS unit to collect the data and mark each landing spot as a waypoint. When you are back in the classroom, use Google Earth to plot out the locations of all of the landings. What patterns do you note in the data?

CHAPTER 8
Aerospace Physiology

Skills List

After completing the activities in this chapter, you should be able to:

- Demonstrate an understanding of how anatomy and physiology work to support daily life and how these same factors limit our performance in the aerospace environment

- Design an experiment that clarifies the limits imposed by human anatomy or physiology or explores the effectiveness of technical, training or procedural solutions for overcoming the limitation

TIP SHEET

The subject of aerospace physiology is very broad, including the full range of body systems and how the normal responses to your environment change in the flight environment. Subjects of inquiry can include the following:

- Human performance (stress, alcohol, carbon monoxide, anemia, temperature, dehydration, drugs, fatigue or jet lag, circadian rhythms, and crew resource management)

- Spatial (dis)orientation (visual, vestibular, otolith)

- Airsickness

- Vestibular system function

- Noise and hearing

- Hypoxia and oxygen equipment or pressurized breathing

- Cabin pressurization and decompression sickness

- Temperature and body function

- Sensory illusions and the visual system (night vision, dark adaptation, night illumination, scanning techniques, flash blindness, night-vision technology)

- G-forces and acceleration (natural and elevated tolerance and endurance), ejection seats

- Reduced or zero equivalent gravity

- Long duration space flight (psychology, hygiene, food, etc.)

- Radiation exposure

- Crash survival

BACKGROUND

The Body in Flight

The human body supports the life functions necessary for our survival on the Earth's surface. When we take our body into an environment that is significantly different from that for which our anatomy and physiology is adapted, our normal responses can be troublesome, debilitating, or deadly. Our bodies are challenged by changes in temperature, pressure, g-forces, oxygenation, and many other factors (Figure 8-1). In this chapter, you will become a subject area expert on one aspect of aerospace physiology and then share your knowledge with a broad audience.

FIGURE 8-1 *A pilot safely ejects from his Thunderbirds F-16 aircraft after maneuvering away from the crowds and just moments before impact.*

Source: Air Force image archive, figure 040130-F-0000C-002

EXERCISE 8.1 AEROSPACE PHYSIOLOGY: TAKING HUMAN PERFORMANCE BEYOND THE NORM

To understand the role that technology and scientific understanding has played in expanding the aerospace envelope of operation, you need to understand the basic limits of human biology and how they change in the aerospace environment. Beyond this, you will need to become familiar with how engineers have used technology to enhance our performance and overcome these limitations.

Objective

Research one aspect of aerospace physiology to identify the normal structure and functions of the human body, the unique conditions imposed by the aerospace environment, and the current and proposed solutions used to overcome these challenges (Figure 8-2).

FIGURE 8-2 *The Holloman air force pilot training centrifuge takes its final spin in 2010 after training thousands of aviators on how to deal with the increased g-forces maneuvering.*

Source: Air Force image archive, figure 101027-F-0295S-182

Materials

- Internet access
- Engineer's notebook
- Presentation materials

Procedure

STEP 1 Obtain one aerospace physiology topic for which you have primary responsibility for becoming the content area expert for your class.

STEP 2 In your engineer's notebook, define the essential nature of this particular subject in terms of who is affected, to what degree they are impaired or endangered, and how this places limits on what humans can accomplish.

STEP 3 Identify, diagram, and define the human anatomical and physiological processes that are the root cause of the aerospace challenge you are studying.

STEP 4 Describe the normal structure and operation of the human anatomy and physiology you have identified. How is this normal function useful to a human carrying out routine daily tasks?

STEP 5 Describe in detail the cause and effect process that makes the anatomy and physiology work to undermine our effort to function in a particular aerospace environment.

STEP 6 Identify and describe the technological, experiential, and procedural solutions that are currently used to overcome the limiting factor that you are studying.

STEP 7 Summarize your findings in a medium that is suitable for public presentation such as a poster session, informational kiosk, museum exhibit, or public speech.

STEP 8 Summarize your findings in written form as a technical report on the current state of our research and development to provide a means to eliminate or reduce the impact of your physiological limiting condition on our performance in the aerospace environment.

STEP 9 Share your public presentation with your class, school, and community.

BACKGROUND

Human Research to Improve Aviation Safety or Performance

Professional engineers are often tasked with developing the tools and techniques to investigate scientific problems. To design an effective experiment, you have to understand the broad strokes of a problem and be able to frame a question that is testable and that provides insight into an essential part of the problem. An experiment may or may not lead directly to a solution for overcoming the problem.

EXERCISE 8.2   **EXPERIMENTAL DESIGN: INVESTIGATING THE LIMITS TO HUMAN PERFORMANCE**

Objective

Design an experiment in which you explore a limitation to human performance in the aerospace environment (Figure 8-3).

FIGURE 8-3 *A Flight Simulator is used to train pilots to deal with the spatial disorientation associated with flying in instrument flight conditions.*

Source: Air Force image archive, figure 090930-F-5306T-914

Materials

- Engineer's notebook
- Graph paper
- Prototyping materials
- Presentation materials

Procedure

STEP 1 ▶ Brainstorm to identify a few essential questions related to the cause-and-effect process that leads to human limitations in one aspect of performing in the aerospace setting.

TIP SHEET

For example, if you believe that airsickness is the direct result of conflicting information from the visual and vestibular systems, then you might pose a question about whether or not blindfolded individuals get airsick.

STEP 2 In your engineer's notebook, list and sketch the equipment, materials, setup, and procedures that could be used to test your essential questions. Take each of your initial questions as far as possible in the time provided.

STEP 3 Select the question that shows the most promise for being an experiment that you could actually complete. Your teacher may or may not tell you that it is okay to ignore limitations of finance or access to equipment for the purposes of this assignment.

STEP 4 Design a complete experiment. Remember to fully explore and describe how the safety of all research subjects will be ensured.

STEP 5 Write a technical report or create a research poster that includes an abstract, objective, background, materials, procedures, expected costs, expected results, and benefits to the aerospace industry if this research were to be completed.

FIGURE 8-4 *A Barany chair is a device that provides a pilot with a first-person experience with spatial disorientation when visual cues are absent. The pilots learn to trust their instruments rather than their physical sensations for navigating in instrument flight conditions.*

Source: Air Force image archive, figure 080423-F-0986R-153

TIP SHEET

Caution: Remember that all human and animal subject research has limitations placed on it by the researching institution, funding agency, and by law (Figure 8-4). Explore whether or not your school district's research policies would or would not allow you to complete your research as you have proposed.

CHAPTER 9
Material Science

Skills List

After completing the activities in this chapter, you should be able to:

- Describe the different characteristics of a material

- Relate material characteristics to the use of a material in an aircraft design

- Comprehend stress, strain, moments of inertia, and Young's modulus of flexibility

- Assemble components that are individually weak or flexible to produce a structure that is both strong and rigid

BACKGROUND

The Materials Science of Aerospace Design

The aerospace industry makes use of many different materials in the construction of structural, functional, and cosmetic components. From the engineer's perspective, the most important strength of materials concerns are the ability of a material to carry a load (stress) with predictable change in shape or position (strain) and a predictable failure mode. In this chapter, you become familiar with materials testing and construction of aerospace structures (Figure 9-1).

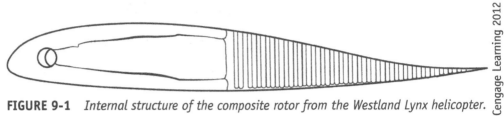

FIGURE 9-1   *Internal structure of the composite rotor from the Westland Lynx helicopter.*

© Cengage Learning 2012

EXERCISE 9.1   COMPOSITE CONSTRUCTION

Objective

Gain firsthand experience of working with composite materials and structures (Figure 9-2).

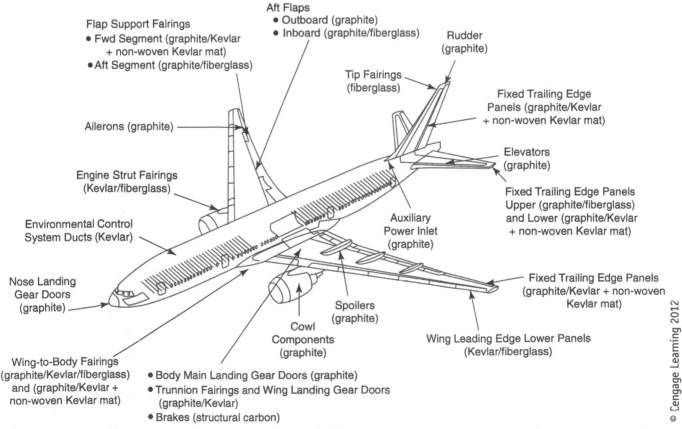

FIGURE 9-2   *Composites can be created using a number of different materials. The specific material selected is matched to the functional demands of the component and the need to be cost effective.*

© Cengage Learning 2012

Materials

- Latex or nitrile gloves
- Safety glasses
- Disposable plastic scrapers/putty knives (3 in. wide)
- Epoxy resin (West System 105 Epoxy Resin)
- Epoxy hardener (West System 206 Slow Hardener)
- Mixing cups
- Mixing sticks
- Scissors
- Fiberglass (bidirectional weave, 2 in. wide)
- Fiberglass (unidirectional sheet, 12 in. wide)
- ¼ or ½ in. thick closed-cell foam
- Roll of 2 mil plastic sheeting to cover tables
- 12 in. shelving
- Heavy weights

Optional:

- ¼ or ½ in. thick pine boards 2 × 12 in.
- ¼ or ½ in. thick oak boards 2 × 12 in.
- Other materials for cores

Procedure

STEP 1 Cover your workspace tables with plastic sheeting, and secure edges so it does not shift.

STEP 2 Create a "layup pressing station" by laying out plastic sheeting that is *not* taped down. Finished composite layups are placed side-by-side on one half of the sheet, and then the sheet is flapped over the layups to be covered and pressed for 24 hours. Use a permanent marker to write labels on the plastic with your team information.

STEP 3 Put on gloves and safety glasses.

STEP 4 Obtain eight strips of bidirectional fiberglass (Bid) and eight strips of unidirectional fiberglass (Uni) each measuring 2 × 12 in.

STEP 5 ▸ Cut two foam cores that measure 2 × 12 in.

STEP 6 ▸ Organize the eight strips and a foam core for one type of fiberglass.

STEP 7 ▸ Follow the mixing directions for epoxy resin and hardener, which is generally equal pumps of both and mix well until uniform in appearance. The "working time" clock starts when mixing starts!

TIP SHEET

Working time refers to how long you have to get the resin/hardener mixture applied to the structures that you are assembling. After the material starts to set, you need to clean up the excess and get a fresh batch mixed. Planning the quantities and time required for a task is a major part of working with composites.

STEP 8 ▸ Pour a sufficient amount of mix onto one strip of fiberglass, and carefully work it into the strip. When finished saturating the strip, squeeze out the excess mix. Use caution to ensure that you don't pull out any fibers.

STEP 9 ▸ Lay the finished strip to the side. If this sample will have a foam core, do this on top of your cut foam core's 2 × 12 in. surface, and press the strip on with the scraper/putty knife.

STEP 10 ▸ In your original working spot, "wet" another strip. When finished, place it on top of the first strip, and press them together. If this is your foam core sample, flip it over when you are done applying the second layer.

STEP 11 ▸ Repeat to wet another two layers, and apply them to your growing layup. If this is a sample without a core, you will wind up with four layers of material completely wetted. If this is a sample with a foam core, you will wind up with two layers of material, a foam core, and then two layers of material.

STEP 12 ▸ Move the completed layup to the press area making certain you save room for the other three layups you are creating.

TIP SHEET

You will find it more effective to put all of the cored samples in one press and the coreless samples into another press so that you can create uniform pressure across each of the samples.

STEP 13 ▸ Repeat the process for the remaining samples. At a minimum, you should have four layups of Uni coreless, Uni cored, Bid coreless, and Bid cored.

STEP 14 ▸ Flip your plastic flap over the samples, place a shelving board over enclosed samples, and place heavy weights uniformly across the top of the shelving board.

STEP 15 ▸ Allow the samples to cure for at least 23 hours before removing them from the press!

STEP 16 ▸ Demold the samples and trim excess hardened epoxy from around the sample with scissors. Wear safety glasses and gloves to prevent injury from flying shards of epoxy.

BACKGROUND

Materials Testing

Many of the primary structural components in an airplane are designed to flex perpendicular to their longitudinal (longest) axis. For example, wing spars are designed to brace the wing so that lift forces are transferred to the fuselage while allowing the wing to flex to absorb variations in the lift force. Engineers balance stiffness and flexibility to achieve their design goals. They accomplish this by selecting materials and shapes that provide the desired outcomes.

EXERCISE 9.2   STRESS AND STRAIN: DEFLECTION TESTING A BEAM

Objective

Simulate a complete wing spar by applying a transverse load to a beam to create a stress in the beam and measure the deflection from neutral position that results (Figure 9-3). Calculate Young's modulus to determine the rigidity/flexibility of each beam.

© Cengage Learning 2012

FIGURE 9-3 *Photo and diagram of the stress testing for the wing panel of the Onex by Sonex Aircraft.*

Source: Courtesy of Sonex Aircraft

Materials

- Engineer's notebook

- Beams of various configuration and materials (2 × 12 in. top surface, various thickness)

- Load bracket (⅛ × 1 × 2 in. aluminum bar)

- Cinder blocks or books
- Ruler
- Force probe
- Ultrasonic range finder
- Data collection software

Optional method:

- Weights
- Stress analyzer

Procedure

Data Collection

STEP 1 Set up the testing station as shown here (Figure 9-4).

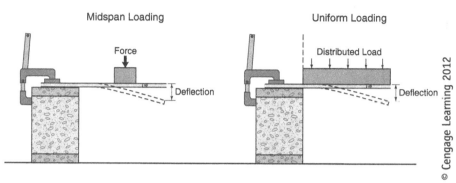

FIGURE 9-4 *How a beam is loaded for testing will impact the deflection that a load will produce. However, the Young's modulus that is calculated for the material in any given orientation should remain the same.*

STEP 2 Use a ruler and calipers to mark the center point of the upper surface of the beam, to draw a line across each beam 2 in. in from each end of the beam and to measure the thickness of the beam.

STEP 3 Place your beam sample across the gap ensuring that at least 2 in. of the beam overlaps onto each cinder block. Place the load bracket across the beam over the center dot.

STEP 4 Zero the force probe, and start data collection.

STEP 5 Use the force probe to press on the center dot of the load bracket until a stress of _____ lbs is being applied to the beam.

TIP SHEET

Apply the force smoothly and gently until it increases to the value desired, and then hold it as steady as you can.

Optional methods: You can also complete this experiment by using a loading bracket that allows you to hang a set amount of weight from the beam; however, your bracket design must not destroy the structure of the beam (Figure 9-5). The same is possible using a stress analyzer, equipment specifically designed to deflect the beam a known amount while recording the stress applied; however, designs for your bracket will vary depending on the equipment used.

STEP 6 Record the deflection in inches that is produced by the deflection force applied.

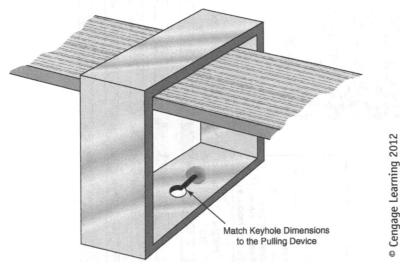

Match Keyhole Dimensions
to the Pulling Device

© Cengage Learning 2012

FIGURE 9-5 *Testing a wing panel or spar in a stress analyzer may require that you make your own adaptor to connect the device's pulling mechanism to the object without drilling holes in the material.*

Data Analysis

For each beam, we need to calculate a moment of inertia and Young's modulus. Moment of Inertia can refer to the tendency for a shape to resist a change in its rotation in space or its resistance to bending or deflection. This last definition is frequently referred to as the Second Moment of Inertia. Two structures that are constructed from the same quantity of material (identical cross-sectional area) can have very different resistance to bending. Consider a lasagna noodle being bent transversely across its width versus across its thickness.

The Second Moment of Inertia (Figures 9-6 and 9-7) for a rectangular beam:

$$I = \frac{b\,h^3}{12}$$

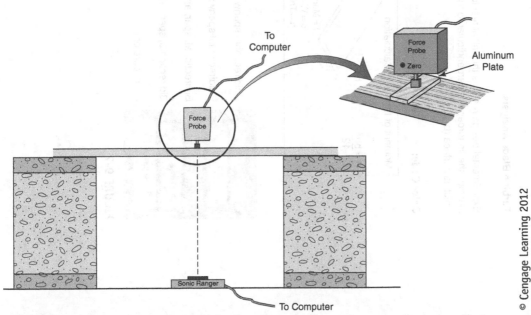

To Computer

Force
Probe

Force
Probe

● Zero

Aluminum
Plate

Sonic Ranger

To Computer

© Cengage Learning 2012

FIGURE 9-6 *Stress testing a spar or wing panel can be done using computer sensors that measure force and location. Using a narrow metal plate that is as wide as the panel ensures that the force is distributed across the width of the beam.*

Composite Materials Testing:

Turbine Blade Analysis:

Stiffness = Load/Deflection

$K = W/y$

Experimental design restricts this initial test loading to conditions in which the lift forces are perpendicular to the turbine blade face which is surface (b).

Color Code: Required Inputs   Calculated Values

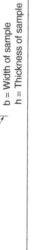

$$y = \frac{wL^3}{48\,E\,I} \qquad E = \frac{wL^3}{48\,I}$$

w = Load (lb)
L = Length of span (in.)
y = Deflection (in.)
I = Moment of inertia (in.^{4})
E = Young's modulus (psi)

Moment of Inertia (I) Calculation

$$I = \frac{bh^3}{12}$$

I = Moment of inertia
b = Width of sample
h = Thickness of sample

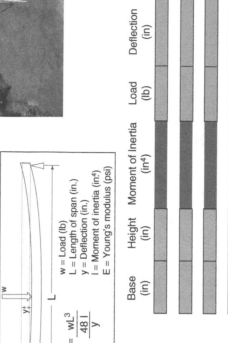

Base (in)	Height (in)	Moment of Inertia (in^4)	Load (lb)	Deflection (in)	Stiffness (lb/in)

Unidirectional spanwise orientation, 4 layers

Unidirectional spanwise orientation, 4 layers, 1" foam core

Bidirectional, span and cross orientation, 4 layers

Bidirectional, span and cross orientation, 4 layers, 1" foam core

Foam core only

© Cengage Learning 2012

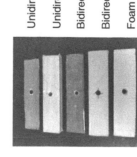

FIGURE 9-7

Deflection Expected:

$$Deflection = \frac{F L^3}{48 E I}$$

If . . .

F = Force Applied (lbs)

L = Distance Between Supports (in.)

E = Modulus of Elasticity or Young's modulus (psi)

I = Second Moment of Inertia (in.^{4})

Then . . .

Deflection is in inches.

STEP 7 Create a spreadsheet using Excel or similar software to summarize and analyze the data collected.

The spreadsheet must include the following:

- A title for the document in row 1
- A data table that includes raw data for the following:

    a. Base width (in.)

    b. Thickness height (in.)

    c. Force applied (lbs)

    d. Length of span (in.)

    e. Deflection produced (in.)

- Data table rows that calculate the value of the following:

    a. Second Moment of Inertia (in.^{4})

    b. Young's modulus (psi)

- Data table row for ranking materials by stiffness

STEP 8 Fill in the spreadsheet with all of your data collected.

STEP 9 Consider what you observed during the testing and the numbers that the spreadsheet are calculating for Young's modulus of elasticity. What does a higher Young's modulus mean about the stiffness of a beam? Rank each of the materials in decreasing order of stiffness so that the most rigid material is ranked number 1.

STEP 10 Write a brief technical report for the main spar of a new aircraft being developed that summarizes the pros and cons of each of the designs tested in this exercise. Consider the impact of how the aircraft is intended to be flown, for example, smooth commercial flight versus aerobatics.

BACKGROUND

Increasing Strength While Reducing Weight

To build an aircraft out of solid components can result in a design that is far too heavy to accomplish the flight performance goals of the project. As such, engineers design to reduce the amount of material required to construct an assembly of components that greatly reduce the weight of the structure while maintaining or enhancing its strength and ability to be inspected and repaired. In this project, you will construct a portion of the leading edge (LE) from a Sonex, Waiex, or Xenos aircraft (www.sonexaircraft.com) to learn the basic metal-working skills required to build a metal aircraft. As a class, you are invited to participate in the online discussion groups and Expercraft Simple Log (online project log book) support network offered by Sonex Aircraft at http://education.sonexaircraft.com/.

EXERCISE 9.3   WEAK MATERIALS, STRONG STRUCTURES, AND THE SONEX PROJECT!

Objective

Construct a small section of the LE and spar from an existing real-world aircraft and consider the strength and weight of the components from which it is assembled (Figure 9-8).

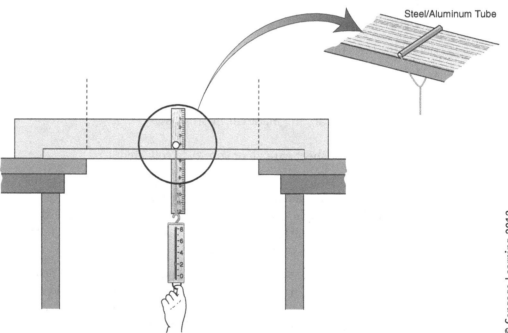

Steel/Aluminum Tube

© Cengage Learning 2012

FIGURE 9-8   *Stress testing a spar or wing panel can also be accomplished with a spring scale and ruler. Using the metal tube that is wider than the panel ensures that the force is distributed across the width of the beam.*

Materials

- Engineer's notebook
- Safety glasses
- Band saw, 5–6 tpi
- Drill press
- Fly cutter, adjustable to 4 in. hole

- Hand pop rivet tool
- Sledgehammer, short handled, 2 lb
- 3/8 in. reversible drill
- ¼ in. drill bit
- #40 drill bit
- #30 drill bit
- #21 drill bit
- Ruler, 12 in. steel, ruled to 64ths
- Rubber mallet
- File, 8 in. half-round, fine tooth
- Sandpaper, 220–440 grit wet-sanding paper (only use dry!)
- Sharpie markers, fine point, blue for centerlines, black for cut lines
- Cleco pliers
- Clecos 3/32 in., (25)
- Clecos 1/8 in., (25)
- Clecos 5/32 in., (8)
- Andy aluminum snips
- C-clamps with swivel pads, (2)
- Vixen file, shared
- Fluting pliers
- Deburring tool
- 3 in. flanging die
- 4 in. flanging die
- Buck bar
- Rivet set
- Wing rib form block and backer

Optional:

- Pneumatic pop riveter
- Compressor, small (if using pneumatic riveter)
- 3M Scotch-Brite Wheel
- Bench grinder (if using Scotch-Brite Wheel)

Supplies:

- Wing skin blank
- Rib blank
- Spar web blank
- Spar cap blank (2)
- Aluminum blind rivets, AAP-42, (14)
- Aluminum blind rivets, AAP-44, (10)
- Solid rivets, AN470/AD5-6, (8)

Procedure

STEP 1 As a class, learn about basic metal-working techniques such as grain, parts layout, cutting, scratches and deburring, bending, drilling, solid riveting, pulled riveting, parts forming, flanging, and fluting.

STEP 2 Obtain the materials required to assemble one 5 in. wide segment of the complete LE segment from a Sonex aircraft.

STEP 3 Discuss the Drawing Tree (SNX-X01) to determine the required workflow for assembling the LE. Note that there are multiple starting points for sequences that flow toward final assembly. Select a starting point for your project (SNX-X09, SNX-X10, SNX-X15, or SNX-X18).

STEP 4 Refer to the engineering drawing for your chosen component, and follow its directions to manufacture the component. It is important to thoroughly understand a manufacturing process before you start to work the material (i.e., "stop, think, act" or "measure twice, and cut, drill, or hammer once"). If in doubt, ask a question.

STEP 5 Inspect your component for accuracy, quality, and completion. Note in your engineer's notebook any remediation steps required to make the part usable.

STEP 6 Have the instructor inspect your component. Carry out any remediation steps that the instructor deems necessary to produce an "airworthy" component.

STEP 7 Repeat steps 3–6 until all four LE components have been manufactured to an airworthy standard.

STEP 8 In sequence, complete the assembly process by working through Main Spar Assembly (SNX-X08), Wing Rib Installation (SNX-X14), and Leading Edge Complete (SNX-X02). At each stage, note that the assembly requires alignment, pilot hole drilling, cleco clamping, final drilling, cleco clamping, disassembly, deburring, reassembly with clecos, and riveting.

STEP 9 After each stage of assembly, inspect and have your instructor inspect the project for accuracy, quality, and completion.

Sonex Leading Edge Project Section

© Cengage Learning 2012

FIGURE 9-9 *Building a small section of the leading edge of the Sonex aircraft teaches most of the basic skills required to build the entire aircraft.*

Source: Courtesy of Sonex Aircraft

RIVETING TOOLS for the MAIN SPAR

The solid rivets used to assemble the main spar can be set using two very simple tools which will eliminate the need for expensive riveting equipment.

The provided bucking bar (A) is contoured so it will not interfere with the spar cap. The bucking bar can be mounted to a heavy surface or secured in a vise so it does not move while the rivets are being set.

The rivet set (B) can be made from a 1/2" hex head cap screw, two nuts, and a large washer. The head of the cap screw should be polished to prevent marring the rivet head and the corners of the hex should be rounded to prevent marring the spar channel.

Two or three firm blows with a 1-1/2 to 2 pound mallet should set the rivet. Make sure the bucking bar is on a very solid surface to prevent bouncing. When setting rivets with this homemade rivet set, the factory head of the rivet will be flattened slightly. This will have no affect on the strength or performance of the rivet.

SONEX AIRCRAFT, LLC

P.O. Box 2521
Oshkosh, WI 54903
Phone (920) 231-8297
fax (920) 426-8333
email: info@sonexaircraft.com
www.sonexaircraft.com

CUTTING the RIB BLANKS

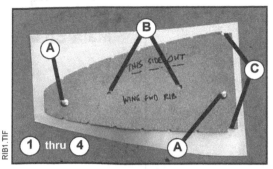

RIB1.TIF ① thru ④

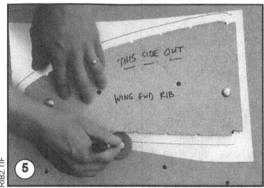

RIB2.TIF ⑤

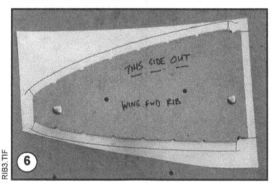

RIB3.TIF ⑥

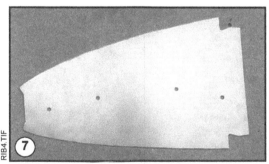

RIB4.TIF ⑦

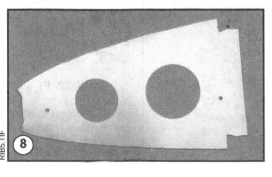

RIB5.TIF ⑧

1. Lay the rib form block on the aluminum with at least 1/2" of aluminum extending out from the form block in all directions.

2. Hold the form block firmly in place and, using the two outer tooling holes (A) as guides, drill through the aluminum sheet and into your workbench with a 1/4" drill bit. After each hole is drilled, insert a 1/4 bolt (A) to hold the form block in position on the aluminum.

3. Using the two inner tooling holes (B) as guides, drill through the aluminum sheet with a 1/4" drill bit.

4. Using the two corner notches (C) as guides, drill through the aluminum sheet with a 1/4" drill bit.

5. Trace around the form block with a washer that measures 1/2" from its outside edge to its inside edge.

6. Complete the outline of the rib blank as shown on the plans and in the accompanying photograph (6).

7. Remove the form block and cut the rib blank from the aluminum sheet.

8. Use the two 1/4" holes in the center of the blank as pilot holes for fly-cutting the two lightning holes.

9. Deburr all of the edges of the rib blank.

10. Form the ribs following the instructions on page SNX-EDU-03.10.

**SNX-EDU-03.9
Cutting the
Rib Blanks**

**Reference
Drawings:
SNX-X15**

SONEX AIRCRAFT, LLC

**P.O. Box 2521
Oshkosh, WI 54903**
Phone (920) 231-8297
fax (920) 426-8333
email: info@sonexaircraft.com
www.sonexaircraft.com

© Cengage Learning 2012

FORMING WING RIBS

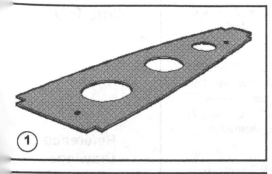

1.

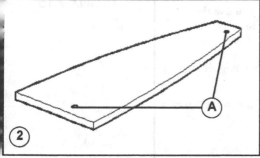

2.

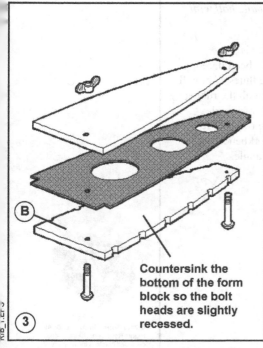

Countersink the bottom of the form block so the bolt heads are slightly recessed.

3.

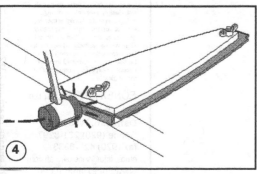

4.

1. Make the flat rib blanks by following the instructions on page SNX-EDU-03.9.

2. Cut a crush plate from 5/8" or 3/4" particle board. The crush plate is the same shape as the form block, but is sized approximately 3/8" smaller all around so the fluting grooves in the form block are fully exposed when the fixture is assembled.

   Locate and drill the fixture holes (A) in the crush plate. This is most easily accomplished by transferring the hole locations from the form block.

3. Obtain two 1/4" bolts and thumb screws. Countersink the fixture holes in the bottom of the form block (B) so the heads of the bolts are below the surface of the form block.

   Assemble the form block, rib blank, and crush plate as shown in the illustration.

4. Bend the flanges of the rib blank to shape by striking them with a plastic tipped mallet until they conform to the outline of the form block as closely as possible.

   Note: The flanges will not conform perfectly to the curves of the form blocks. The process of fluting the flanges in step 7 will make the flanges conform to the proper shape.

SNX-EDU-03.10
Forming Wing Ribs

Reference Drawings: SNX-X15

SONEX AIRCRAFT, LLC

P.O. Box 2521
Oshkosh, WI 54903
Phone (920) 231-8297
fax (920) 426-8333
email: info@sonexaircraft.com
www.sonexaircraft.com

© Cengage Learning 2012

FORMING WING RIBS

SNX-EDU-03.11
**Forming
Wing Ribs**

**Reference
Drawings:**
SNX-X15

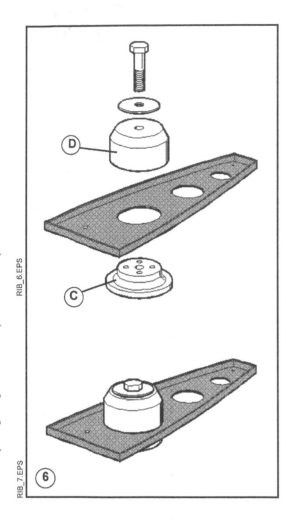

RIB_6.EPS

RIB_7.EPS

(6)

5. After the flanges have been formed, mark the location of each flute on the rib blank.

6. Remove the rib from the form block and flare each lightening hole by sliding the male half of the die (C) through the lightening hole and tightening the female half of the die (D) firmly against it. Refer to the illustration.

Important: There are two different sized lightening holes. Make sure you use the correct sized die for each hole.

Note: An alternate method of forming the flares is to:

a. *secure the male half of the die to the top of the workbench*

b. *place the rib over the die*

c. *place the female half of the die over the male half and strike the female half with the plastic tipped mallet.*

7. After the lightening holes have been flared, use a fluting pliers to form the flutes. You will also have to work the flanges with the rubber mallet to achieve the proper flange angle. Your goal in fluting each wing rib is to get it to lie perfectly flat on your workbench with the flanges bent to the proper angle.

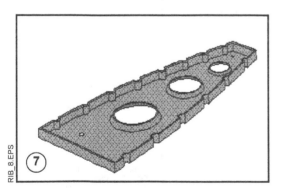

RIB_8.EPS

(7)

SONEX AIRCRAFT, LLC

**P.O. Box 2521
Oshkosh, WI 54903**
Phone (920) 231-8297
fax (920) 426-8333
email: info@sonexaircraft.com
www.sonexaircraft.com

© Cengage Learning 2012

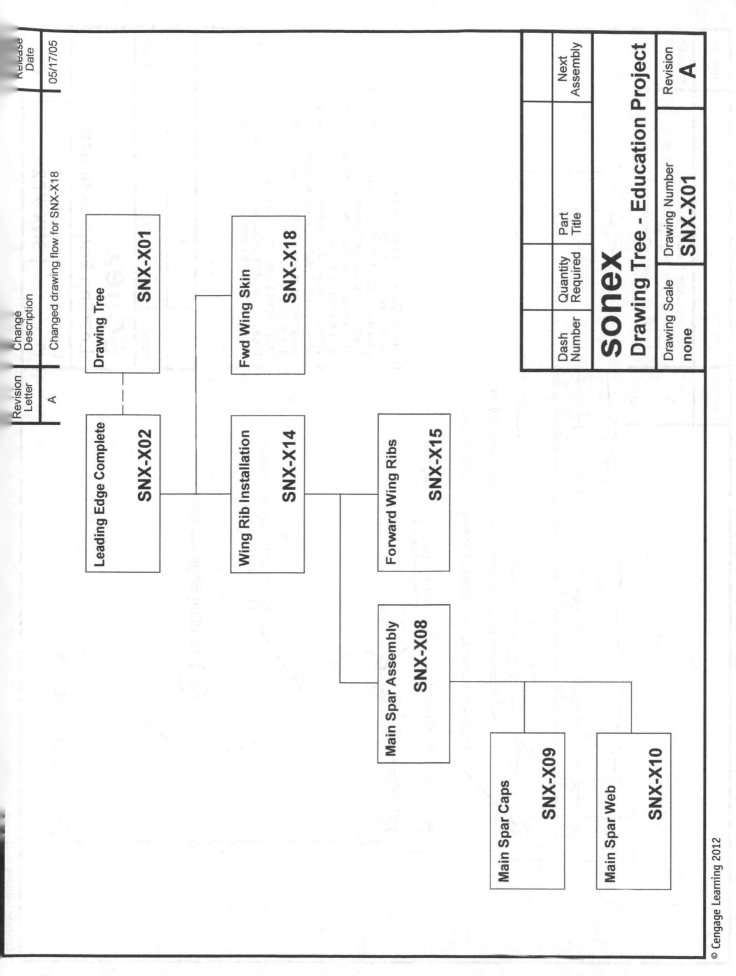

© Cengage Learning 2012

Revision Letter	Change Description	Release Date
A	#30 drill in step 5 was #21.	06/04/01
B	7/16" dimension for positioning leading edge over spar cap was 11/16".	05/05/03
C	AAP Rivets were CCP Rivets.	01/18/05
D	Removed one rib.	05/17/05

General Construction Order

1. Clamp skin in place on top and bottom spar cap.
2. Pilot drill 3/32" (#40) holes through spar caps using pilot holes in wing skin as guides.
3. Align rib so "blue lines" on rib flanges show through pilot holes in wing skin.
4. Working forward from the spar to the leading edge, pilot drill 3/32" (#40) holes in rib using pilot holes in wing skin as guides. Install a cleco in each hole after it is drilled.
5. Drill holes to final size with a #30 drill.
6. Disassemble and debur both sides of all holes.
7. Re-assemble with clecos and install rivets.

Attach skin to rib using Cherry AAP-42 Blind Rivets, Qty. 14.

Attach skin to spar cap using Cherry AAP-44 Blind Rivets, 10 total.

SNX-X18-02, Forward Wing Skin
1 Required on -01

SNX-X14-01, Wing Rib Installation
1 Required on -01

7/16" [11.1mm] Typical

(-01) **Leading Edge Complete**

	Dash Number	Quantity Required	Part Title		Next Assembly

sonex
Leading Edge Complete

Drawing Scale	Drawing Number	Revision
1/2	SNX-X02	D

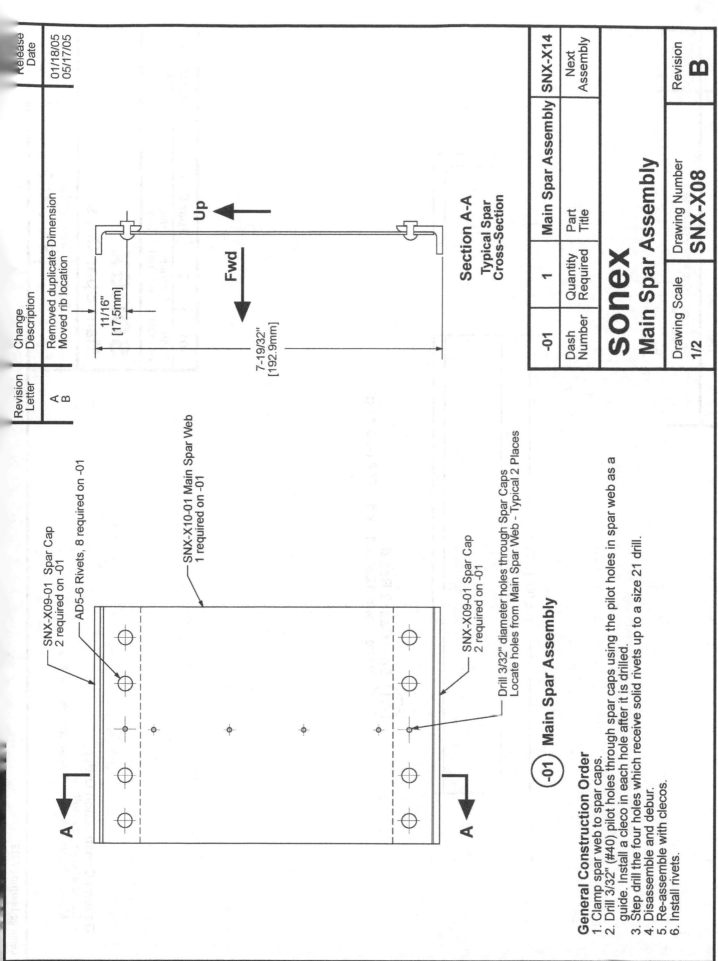

Section A-A
Typical Spar Cross-Section

Up

Fwd

11/16"
[17.5mm]

7-19/32"
[192.9mm]

SNX-X09-01 Spar Cap
2 required on -01

AD5-6 Rivets, 8 required on -01

SNX-X10-01 Main Spar Web
1 required on -01

SNX-X09-01 Spar Cap
2 required on -01

Drill 3/32" diameter holes through Spar Caps
Locate holes from Main Spar Web - Typical 2 Places

A

A

(-01) Main Spar Assembly

General Construction Order
1. Clamp spar web to spar caps.
2. Drill 3/32" (#40) pilot holes through spar caps using the pilot holes in spar web as a guide. Install a cleco in each hole after it is drilled.
3. Step drill the four holes which receive solid rivets up to a size 21 drill.
4. Disassemble and debur.
5. Re-assemble with clecos.
6. Install rivets.

Release Date	Revision Letter	Change Description
01/18/05	A	Removed duplicate Dimension
05/17/05	B	Moved rib location

		Main Spar Assembly	SNX-X14
-01	1	Part Title	Next Assembly
Dash Number	Quantity Required		

sonex
Main Spar Assembly

Drawing Scale	Drawing Number	Revision
1/2	SNX-X08	B

© Cengage Learning 2012

Revision Letter	Change Description	Release Date
A	3/4" was 5/8" to inside of flange.	01/29/03
B	5.00" was 6.00"	05/17/05

3/4"
[19.0mm]

5.00" [127mm]

⊖ -01   Spar Cap, 2 Req'd.
Make from .125" x 1" x 1" 6061-T6 Angle

-01	2	Spar Cap	SNX-X08
Dash Number	Quantity Required	Part Title	Next Assembly

sonex
Main Spar Caps

Drawing Scale	Drawing Number	Revision
FULL	SNX-X09	B

General Construction Order
1. Cut spar caps to length.
2. Mark, trim, and finish each spar cap per drawing.

© Cengage Learning 2012

Revision Letter	Change Description	Release Date
A	168.6mm dimension was 117.9mm	07/12/02
B	5.00" was 6.00"	05/17/05

Main Spar Web
.025 Thick 6061-T6 Aluminum Sheet

(-01)

General Construction Order
1. Debur edges of blank.
2. Mark hole locations.
3. Label top of spar web.
4. Drill 3/32" (#40) pilot holes.

3/32" Pilot Holes, 14 Places

TOP

SNX-X08	Main Spar Web	
Next Assembly	Part Title	
-01	1	Main Spar Web
Dash Number	Quantity Required	

sonex
Main Spar Web

Drawing Scale	Drawing Number	Revision
3/4	SNX-X10	B

2-1/2" [63.5mm]

4.00" [101.6mm]
4 Holes Spaced 1.00" On Center

5.00" [127mm]

1/2" [12.7mm]

27/64" [10.7mm]

1-5/64" [27.4mm]

2-23/32" [69mm]

4-11/32" [110.3mm]

5-63/64" [152mm]

6-41/64" [168.6mm]

7-1/16" [179.4mm]

Revision Letter	Change Description	Release Date
A	Hidden line in rear view changed to solid line.	05/05/03
B	AAP rivets were CCP rivets	01/18/05
C	One rib removed	05/17/05

AAP-44 Rivets
2 required on –01

AAP-42 Rivets
4 required on –01

SNX-X15-01 Front Rib
1 required on –01

SNX-X10-01 Spar Web
1 required on –01

1/4" [6.4mm]

Position forward rib so wing skin will lie smoothly across rib and spar caps.

(-01) **Wing Rib Installation**

General Construction Order

1. Clamp rib in place with "blue line" on rear rib flange visible through pilot holes in main spar, and top and bottom flanges properly positioned for smooth leading edge skin installation.
2. Drill 3/32"(#40) pilot holes through rear rib flange using holes in main spar as guides.
3. Step drill 3/32 holes to #30 drill size.
4. Disassemble and debur holes.
5. Re-assemble and rivet rib in place.

	Wing Rib Installation	SNX-X02
1	Part Title	Next Assembly
–01	Quantity Required	
Dash Number		

sonex
Wing Rib Installation

Drawing Scale	Drawing Number	Revision
1/2	SNX-X14	C

© Cengage Learning 2012

4" Lightening Hole.
Locate from 1/4" pilot hole in form block.

Spar cap clearance notch (2 Places).

1/4" Diameter tooling hole.
Typical 2 places.
Locate from form block.

Outside dotted line represents outline of rib blank
before forming. Trace form block with proper sized
washer to achieve 1/2' flange. Typical 3 places.

3" Lightening Hole.
Locate from 1/4" pilot hole
in form block.

Forward Wing Rib
.025 Thick 6061-T6 Aluminum Sheet
-01

-01	1	Forward Wing Rib	SNX-X14
Dash Number	Quantity Required	Part Title	Next Assembly

sonex
Forward Wing Rib

Drawing Scale	Drawing Number	Revision
1/2	**SNX-X15**	

General Construction Order
1. Follow the directions in the builder's manual, pages SNX-EDU-03.9 thru SNX-EDU-03.11 to build the rib.

© Cengage Learning 2012

Revision Letter	Change Description	Release Date
A	5.00" Width was 6.00"	05/17/05

Forward Wing Skin Layout
.025 Thick 6061-T6 Aluminum Sheet

Leading Edge Bend Reference Line

3/32" Pilot Holes, 24 Places

BOTTOM

2.00" [50.8mm]

15-1/8" [384.2mm]

11-3/4" [298.5mm]

10.00" [254mm]
5 Holes Spaced on 2" Centers

3/8" [9.5mm]

2-1/2" [63.5mm]

31.00" [787.4mm]

TOP

12.00" [304.8mm]
6 Holes Spaced on 2" Centers

2.00" [50.8mm]

3/8" [9.5mm]

4.00" [101.6mm]

5.00" [127mm]

1/2" [12.7mm]

4 Holes Spaced on 1.00" Centers (Typical)

-01

General Construction Order
1. Debur edges of blank.
2. Mark hole locations.
3. Mark leading edge bend reference line.
4. Label top of leading edge.
5. Drill 3/32" (#40) pilot holes.
6. Form leading edge profile.

Forward Wing Skin
Make from X18-01 Forward Wing Skin Layout

-02

Leading Edge Bend Reference Line

Dash Number	Quantity Required	Part Title	Next Assembly
-02	1	Forward Wing Skin	SNX-X02

sonex
Forward Wing Skin

Drawing Scale	Drawing Number	Revision
1/4	SNX-X18	A

© Cengage Learning 2012

CHAPTER 10
Remote System Design

Skills List

After completing the activities in this chapter, you should be able to:

- Understand the wide range of applications that make use of robotics and industrial/task automation

- Make effective use of pseudocode to provide instructions to complete a task

- Comprehend the challenges of telemetry and the need for autonomous operations

- Develop a task challenge for a robot to accomplish

- Understand the range of robotics competitions available to high school students

BACKGROUND

Robotics and Industrial Automation

As technological solutions provide the means to overcome an increasing number of our challenges, an understanding of robotics and automation is becoming an essential component of technical literacy. To understand this discipline, we need to understand what robots can and cannot do (Figure 10-1). This means we must become familiar with the tasks they already perform, how they are programmed to carry out their tasks, why automation is an essential operational mode for many applications, and how to design a task to be performed by a robotic or automated mechanism.

FIGURE 10-1 *The Space Shuttle's long robotic arm allows the crew to manipulate objects during a mission.*
Source: Courtesy of NASA Archive

EXERCISE 10.1   ROBOTS IN SPACE . . . AND EVERYWHERE ELSE!

Objective

Explore the wide range of robotics applications (Figure 10-2).

FIGURE 10-2 *The Mars rover*
Source: Courtesy of NASA NIX Archive

Materials

- Engineers notebook
- Internet access

Procedure

STEP 1 Explore one of the following segments of the use of robotics and automation:

- Industry
- Space exploration
- Military
- Home automation
- Hostile environments
- Search and rescue
- Entertainment
- Hospitals
- Human companions
- Agriculture

STEP 2 Within your subject, identify at least six different applications that make use of robotic or automated equipment to complete a task. Use the list provided here to structure your recorded information for each application.

- Identify the task completed
- How this task is performed by humans
- Risks and/or deficiencies of having human workers involved
- Images or diagrams
- Dimensions
- Operational description
- Pricing

STEP 3 Meet with the other researchers within your subject. Use a round-robin technique to summarize your findings and those of the group. Each person summarizes one application within the subject while the other members listen. When you are a listener, use your engineer's notebook to note the questions that you have for each application. Rotate to the left until everyone has reported on all of their applications.

STEP 4 Return to each application and round-robin again to ask all of the questions that you have about the application.

STEP 5 As a group, identify the three most interesting applications that were identified as well as the most common applications. These are your four examples for the use of robotics and automation in your subject area.

STEP 6 Organize the research group so that each person has primary responsibility for at least one application and secondary responsibility for at least one more application. Ensure that all four of your examples are covered by a primary and a secondary investigator.

STEP 7 Turn in a list to the instructor documenting the research responsibilities of everyone in your group.

STEP 8 The primary and secondary investigators should carry out research to answer all of the questions that were raised by the group and then meet together to organize the information they discovered.

STEP 9 As a group, create a public presentation that summarizes your four examples.

STEP 10 Give your presentations.

BACKGROUND

Developing the Logic behind a Program

The technique of pseudocoding is used by programmers as a method for exploring and debugging the logic of a sequence of operations using language written for human comprehension rather than execution by a machine. Thus pseudocoding allows programming to start without having to deal with the other potential problems involved with actual software or hardware. This is useful for discussing the logic of a program with others, optimizing the sequence of operations, minimizing steps required to complete a task, and structuring the overall code.

EXERCISE 10.2 PSEUDOCODING: TELL IT LIKE IT IS

Objective

Use pseudocode to provide instructions for another human to complete a task (refer to Example 10.1).

Example 10.1

Pseudocode for calculating an average grade for a class:

1. Find the sum of all points earned.

2. Find the sum of all points possible.

3. Divide the total points earned by total points possible, and multiply by 100.

4. Record the result as a percentage.

Basics of pseudocode:

- Pseudocode includes many operational words such as add, sum, divide, initialize, compute, process, print, output, and so on.

- Pseudocode uses numbers to indicate the sequence of operations.

- Pseudocode can indicate looping or decisions with phrases such as if . . . then, if . . . else, do while . . . return, do until . . . end do, and so on.

Materials

- Engineer's notebook
- Paper
- Pen or pencil
- Miscellaneous materials

Procedure

STEP 1 Your teacher has set up a complex but common task.

STEP 2 Use pseudocode to provide a "programming" solution for this automation challenge. You can assume that your "robot" is as capable as a typical human.

STEP 3 Exchange your code with others.

STEP 4 Partner with an execution monitor to ensure that you follow the pseudocode step by step.

STEP 5 Execute the pseudocode that you have been given, and then exchange roles with your execution monitor.

BACKGROUND

Communicating with a Robot

Telemetry is the process of encoding, broadcasting, receiving, and decoding information. In many circumstances, it is desirable to have the information gathered by the sensors of a robot or the data that it calculates transmitted back to human operators. During space operations, there can be significant delays between data transmission and reception due to the distances involved.

EXERCISE 10.3 TELEMETRY AND ANTONYMOUS OPERATIONS

Objective

Use an information encoding, broadcasting, receiving, decoding process to simulate telemetry from a robot (Figure 10-3).

FIGURE 10-3 *Radio telescopes such as the Very Large Array (VLA) are used to observe the universe at very long wavelengths. At extreme distances, antennas similar to these become necessary for maintaining contact with deep-space probes.*

Source: Image courtesy of NRAO/AUI/NSF

Materials

- Engineer's notebook
- Note cards
- Pen or pencil
- Miscellaneous materials

Procedure

STEP 1 ▶ You will be partnered in a group with a mission controller, a telemetry officer, a safety officer, and a robot.

STEP 2 ▶ The mission controller will be allowed to enter into a room that contains a complex task for the robot to complete. The mission controller will have eight minutes to observe, sense, and encode the nature of the task.

STEP 3 Mission controllers will be placed in one room while robots are in the task room with a secure blindfold on. A safety officer will accompany the robot at *all* times.

STEP 4 The telemetry officer will enter the mission controller room.

1. Obtain a note card from the mission controller with one step of the challenge solution encoded on it.

2. Walk into the task room.

3. Announce the challenge step.

4. Wait for the step to be executed.

5. Return to mission control.

6. Return the note card to the mission controller.

7. Repeat until time expires or all robots have achieved the task.

STEP 5 Debrief the mission by reviewing with the entire group the task challenge, the pseudocode instructions of the most successful robot team, the pseudocode instructions of the least successful robot team, and the relationship of this task to a real mission.

BACKGROUND

Optimizing Tasks for Robot Workers

When tasks are designed for robots, they take into account the natural motions of a robot, simplifying them to a single motion, or placing objects in a consistent location. In this exercise, you will design a task for completion by a robot.

EXERCISE 10.4 TASK DESIGN FOR AUTOMATION OPTIMIZATION

Objective

Design a task to be completed by an automated system or robot.

Materials

- Engineer's notebook
- Internet access
- (Optional) Prototyping materials

Procedure

STEP 1 Select a task to be completed (Figure 10-4).

Ignis

Prometheus

© Cengage Learning 2012

FIGURE 10-4 *The BadgerBOTS Robotics team from Madison, Wisconsin (#1306), built Prometheus and Ignis to compete in the 2011 FIRST Robotics Competition entitled LogoMotion.*

Source: Photo by Ben Senson

1. Place a box, fill it with the correct amount of product, close and seal the box, move the package to the side, and repeat.

2. Sort and count prescription medications on a schedule for a patient, monitoring the need for refilling storage.

3. Sort packages or envelopes by color, bar code, zip code, and so on.

4. Assemble a simple vehicle out of LEGOs . . . fully automatically.

5. Identify and describe a task of your own.

STEP 2 Carry out design research from readings, Web resources, expert interviews, or observations of similar solutions.

STEP 3 Brainstorm solutions for your task considering materials, resources, tools, skills, mechanisms, overall layout, flow of materials, and so on. There are no bad ideas, so keep them all in your engineer's notebook (Figure 10-5).

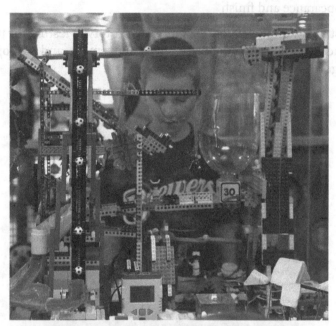

FIGURE 10-5 *An interactive exhibit was created by members of the BadgerBOTS Robotics FIRST LEGO teams for the Madison Children's Museum. The exhibit contains six interactive modules at any given time. All make use of the NXT LEGO robotics components.*

Source: Photo by BadgerBOTS Robotics

STEP 4 Partner with two other engineers. Share your preliminary design concepts with the group. The two nonpresenting engineers should actively listen and identify at least two but not more than four areas that require clarification in the design presented.

STEP 5 Each engineer will draft a memo to each of the two partners communicating in writing the areas of design that require refinement, including questions to be answered, issues to be resolved, limitations to be overcome, and concepts to be proven.

STEP 6 Carry out additional research as required to address the concerns of your partnering engineers.

STEP 7 ▶ Write a design brief that details the following:

- Task to be accomplished
- Dimensions
- Constraints
- Materials (type and suitability)
- Required functions
  - o Movement
  - o Manipulation
  - o Sensors
  - o Forces
  - o Accelerations
  - o Safety
  - o Appearance and finish

STEP 8 ▶ Test your design proposal/solution by prototype testing for proof of concept and evaluation of size, shape, movement, and task optimization (Figure 10-6).

© Cengage Learning 2012

FIGURE 10-6 *A LEGO NXT robot charges forward during a summer camp session to challenge a mini-figure knight at jousting.*

Source: Photo by Ben Senson

STEP 9 ▶ Create a design presentation summarizing the project.

Competing with Your Knowledge

Learning becomes meaningful when there is an application for the skills and knowledge gained from an experience (Figure 10-7). Robotics competitions and related activities provide real-world applications for the knowledge that you have gained. The wide range of opportunities allows people of different levels of proficiency access to resources, and time to commit to the endeavor of exploring robotics beyond the typical classroom activities.

FIGURE 10-7 *Programming robots can be done using visual languages such as LabView-based NXT-G or many text-based languages. Here, two summer camp programmers are modifying their codes so that their robot will move a pirate ship during a challenge.*

Source: Photo by Ben Senson

EXERCISE 10.5 ROBOTICS COMPETITIONS AND ACTIVITIES

Objective

Become familiar with a large number of first-person, experiential learning opportunities in the field of robotics and remote operations (Figure 10-8).

FIGURE 10-8 *Participating in robotics is an enjoyable activity. Opportunities range from hobby activities at home to summer camps and workshops, to competitive experiences such as those offered through the FIRST robotics programs.*

Source: Photo by Ben Senson

© Cengage Learning 2012

Materials

- Engineers notebook
- Internet access

Procedure

STEP 1 Explore one segment of robotics and remote operations competitions and events:

- Junior FIRST LEGO League
- FIRST LEGO League
- FIRST Tech Challenge
- FIRST Robotics Competition
- Vex Robotics Competition
- National Robotics Competition (SME Education Foundation)
- BEST Robotics
- BotBall
- Lunar XPrize
- RoboSumo
- RoboCup Soccer
- Marine Advanced Technology Education ROV
- Navy Robotic Submarine Competition
- SeaPerch
- Others

STEP 2 Within your subject, carry out individual research to identify the essential elements of the competition or event:

- Official name of competition or event
- Web site
- Schedule of events
- Requirements for participation
- Rules
- Costs of participation
- Examples of prior and current challenges
- Nearest location of already-participating groups
- Feasibility of local participation

STEP 3 Meet with the other researchers within your subject. Use a round-robin technique to summarize your findings and those of the group. Each person summarizes one aspect within the subject while the other members listen. When you are a listener, use your engineer's notebook to note the questions that you have for each aspect. Rotate to the left until everyone has reported on all of their identified aspects of the competition or event.

STEP 4 Return to each presenter and round-robin again to ask all of the questions that you have about the research.

STEP 5 As a group, identify the most interesting elements within each aspect of the competition or event that were identified by the overall group.

STEP 6 Organize the research group so that each person has primary responsibility and secondary responsibility for an equal number of the primary competition aspects. Ensure that all competition aspects are covered by a primary and a secondary investigator.

STEP 7 Turn in a list to the instructor documenting the research responsibilities of everyone in your group.

STEP 8 The primary and secondary investigators should carry out research to answer all of the questions that were raised by the group and then meet together to organize the information they discovered.

STEP 9 As a group, create a public presentation that summarizes the competition.

STEP 10 Give your presentations.

Continued from page

SIGNATURE: DATE:

WITNESSED BY: DATE:

PROPRIETARY INFORMATION

Continued from page

Continued from page

SIGNATURE:

DATE:

WITNESSED BY:

DATE:

PROPRIETARY INFORMATION

Continued from page

Continued from page

SIGNATURE:

DATE:

WITNESSED BY:

DATE:

PROPRIETARY INFORMATION

Continued from page

Continued from page

SIGNATURE:

DATE:

WITNESSED BY:

DATE:

PROPRIETARY INFORMATION

Continued from page

Continued from page

SIGNATURE: DATE:

WITNESSED BY: DATE:

PROPRIETARY INFORMATION

Continued from page

Continued from page

SIGNATURE:

DATE:

WITNESSED BY:

DATE:

PROPRIETARY INFORMATION

continued from page

Continued from page

SIGNATURE: DATE:

WITNESSED BY: DATE:

PROPRIETARY INFORMATION

Continued from page

Continued from page

SIGNATURE:

DATE:

WITNESSED BY:

DATE:

PROPRIETARY INFORMATION

continued from page

SIGNATURE:

DATE:

WITNESSED BY:

DATE:

Continued from page

PROPRIETARY INFORMATION

Continued from page

Continued from page

SIGNATURE: DATE:

WITNESSED BY: DATE:

PROPRIETARY INFORMATION